ANDERS ERICSSON

Professor Anders Ericsson is the world's reigning expert on expertise. His research into what makes ordinary people achieve the extraordinary was the inspiration for the 10,000-hours rule – the popular theory that 10,000 hours of any type of practice will allow an individual to excel in any field. More recently, he has investigated how a particular type of extended practice leads to exceptional performance.

Ericsson's research has been widely cited in newspapers and magazines worldwide. He has worked with major international organisations, as well as Oxford, Stanford and Harvard medical schools, teachers and educational researchers, professional sports teams, and military groups.

ROBERT POOL

Robert Pool is a science writer who has worked at some of the world's most prestigious science publications, including *Science* and *Nature*, and his writing has appeared in many others. He is the author of three previous books.

ANDERS ERICSSON AND
ROBERT POOL

Peak

How All of Us Can Achieve Extraordinary Things

VINTAGE

1 3 5 7 9 10 8 6 4 2

Vintage
20 Vauxhall Bridge Road,
London SW1V 2SA

Vintage is part of the Penguin Random House group of companies whose addresses can be found at global.penguinrandomhouse.com.

First published in Vintage in 2017
First published in hardback by The Bodley Head in 2016
First published in the United States
by Houghton Mifflin Harcourt in 2016

penguin.co.uk/vintage

A CIP catalogue record for this book is available from the British Library

ISBN 9780099598473

Book design by Chrissy Kurpeski

Printed and bound in Great Britain by Clays Ltd, St Ives plc

Penguin Random House is committed to a sustainable future for our business, our readers and our planet. This book is made from Forest Stewardship Council® certified paper.

To my wife, Natalie, for facilitating and encouraging my efforts to keep pushing beyond my current level of understanding of expert performance and getting closer to the peak.

— A. E.

To my soul mate and my muse, Deanne,
who taught me
much of what I know about writing,
most of what I know about living,
and all of what I know about love.

— R. P.

Contents

Authors' Note

This book is the product of a collaboration between two people, a psychological scientist and a science writer. We began talking regularly about the subject — expert performers and "deliberate practice" — more than a decade ago and began serious work on the book more than five years ago. During that time the book grew in the give-and-take between the two of us to the point that it is now difficult even for us to tell exactly who is responsible for which piece of it. What we do know is that it is a much better — and different — book than either of us would have produced alone.

However, while the book is a collaboration, the story that it tells is the story of just one of us (Ericsson), who has spent his adult life studying the secrets of extraordinary performers. Thus, we chose to write the book from his point of view, and the "I" in the text should be understood as referring to him. Nonetheless, the book is our joint effort to describe this exceptionally important topic and its implications.

Anders Ericsson
Robert Pool
October 2015

Introduction: The Gift

WHY ARE SOME PEOPLE so amazingly good at what they do? Anywhere you look, from competitive sports and musical performance to science, medicine, and business, there always seem to be a few exceptional sorts who dazzle us with what they can do and how well they do it. And when we are confronted with such an exceptional person, we naturally tend to conclude that this person was born with something a little extra. "He is so gifted," we say, or, "She has a real gift."

But is that really so? For more than thirty years I have studied these people, the special ones who stand out as experts in their fields — athletes, musicians, chess players, doctors, salespeople, teachers, and more. I have delved into the nuts and bolts of what they do and how they do it. I have observed, interviewed, and tested them. I have explored the psychology, the physiology, and the neuroanatomy of these extraordinary people. And over time I've come to understand that, yes, these people do have an extraordinary gift, which lies at the heart of their capabilities. But it is not the gift that people usually assume it to be, and it is even more powerful than we imagine. Most importantly, it is a gift that every one of us is born with and can, with the right approach, take advantage of.

THE LESSON OF PERFECT PITCH

The year is 1763, and a young Wolfgang Amadeus Mozart is about to embark on a tour around Europe that will jump-start the Mozart legend. Just seven years old and barely tall enough to see over the top of a harpsichord, he captivates audiences in his hometown of Salzburg with his skill on the violin and various keyboard instruments. He plays with a facility that seems impossible to believe in someone so young. But Mozart has another trick up his sleeve that is, if anything, even more surprising to the people of his era. We know about this talent because it was described in a rather breathless letter to the editor about the young Mozart that was published in a newspaper in Augsburg, Mozart's father's hometown, shortly before Mozart and his family left Salzburg for their tour.

The letter writer reported that when the young Mozart heard a note played on a musical instrument — any note — he could immediately identify exactly which note it was: the A-sharp in the second octave above middle C, perhaps, or the E-flat below middle C. Mozart could do this even if he was in another room and could not see the instrument being played, and he could do it not just for the violin and fortepiano but for every instrument he heard — and Mozart's father, as a composer and music teacher, had nearly every imaginable musical instrument in his house. Nor was it just musical instruments. The boy could identify the notes produced by anything that was sufficiently musical — the chime of a clock, the toll of a bell, the ah-choo of a sneeze. It was an ability that most adult musicians of the time, even the most experienced, could not match, and it seemed, even more than

Mozart's skill on keyboard and violin, to be an example of the mysterious gifts that this young prodigy had been born with.

That ability is not quite so mysterious to us today, of course. We know a good deal more about it now than 250 years ago, and most people today have at least heard of it. The technical term is "absolute pitch," although it is better known as "perfect pitch," and it is exceptionally rare — only about one in every ten thousand people has it. It is much less rare among world-class musicians than among the rest of us, but even among virtuosos it is far from normal: Beethoven is thought to have had it; Brahms did not. Vladimir Horowitz had it; Igor Stravinsky did not. Frank Sinatra had it; Miles Davis did not.

It would seem, in short, to be a perfect example of an innate talent that a few lucky people are born with and most are not. Indeed, this is what was widely believed for at least two hundred years. But over the past few decades a very different understanding of perfect pitch has emerged, one that points to an equally different vision of the sorts of gifts that life has to offer.

The first hint emerged with the observation that the only people who had received this "gift" had also received some sort of musical training early in their childhood. In particular, a good deal of research has shown that nearly everyone with perfect pitch began musical training at a very young age — generally around three to five years old. But if perfect pitch is an innate ability, something that you are either born with or not, then it shouldn't make any difference whether you receive music training as a child. All that should matter is that you get enough musical training — at any time in your life — to learn the names of the notes.

The next clue appeared when researchers noticed that perfect pitch is much more common among people who speak a tonal language, such as Mandarin, Vietnamese, and several other Asian tongues, in which the meaning of words is dependent on their pitch. If perfect pitch is indeed a genetic gift, then the only way that the tonal-language connection would make sense would be if people of Asian ancestry are more likely to have genes for perfect pitch than people whose ancestors came from elsewhere, such as Europe or Africa. But that is something that is easy to test for. You just recruit a number of people of Asian ancestry who grew up speaking English or some other nontonal language and see if they are more likely to have perfect pitch. That research has been done, and it turns out that people of Asian heritage who don't grow up speaking a tonal language are no more likely than people of other ethnicities to have perfect pitch. So it's not the Asian genetic heritage but rather learning a tonal language that makes having perfect pitch more likely.

Up until a few years ago, this was pretty much what we knew: Studying music as a child was thought to be essential to having perfect pitch, and growing up speaking a tonal language increased your odds of having perfect pitch. Scientists could not say with certainty whether perfect pitch was an innate talent, but they knew that if it was a gift, it was a gift that only appeared among those people who had received some training in pitch in childhood. In other words, it would have to be some sort of "use it or lose it" gift. Even the lucky few people who are born with a gift for perfect pitch would have to do something — in particular, some sort of musical training while young — to develop it.

We now know that this isn't the case, either. The true character of perfect pitch was revealed in 2014, thanks to a beautiful experiment carried out at the Ichionkai Music School in Tokyo and reported in the scientific journal *Psychology of Music.* The Japanese psychologist Ayako Sakakibara recruited twenty-four children between the ages of two and six and put them through a months-long training course designed to teach them to identify, simply by their sound, various chords played on the piano. The chords were all major chords with three notes, such as a C-major chord with middle C and the E and G notes immediately above middle C. The children were given four or five short training sessions per day, each lasting just a few minutes, and each child continued training until he or she could identify all fourteen of the target chords that Sakakibara had selected. Some of the children completed the training in less than a year, while others took as long as a year and a half. Then, once a child had learned to identify the fourteen chords, Sakakibara tested that child to see if he or she could correctly name individual notes. After completing training every one of the children in the study had developed perfect pitch and could identify individual notes played on the piano.

This is an astonishing result. While in normal circumstances only one in every ten thousand people develops perfect pitch, every single one of Sakakibara's students did. The clear implication is that perfect pitch, far from being a gift bestowed upon only a lucky few, is an ability that pretty much anyone can develop with the right exposure and training. The study has completely rewritten our understanding of perfect pitch.

So what about Mozart's perfect pitch? A little investigation into his background gives us a pretty good idea of what happened. Wolfgang's father, Leopold Mozart, was a moderately talented violinist and composer who had never had the degree of success he desired, so he set out to turn his children into the sort of musicians he himself had always wanted to be. He began with Mozart's older sister, Maria Anna, who by the time she was eleven was described by contemporaries as playing the piano and harpsichord as well as professional adult musicians. The elder Mozart — who wrote the first training book for children's musical development — began working with Wolfgang at an even younger age than he had started with Maria Anna. By the time Wolfgang was four, his father was working with him full time — on the violin, the keyboard, and more. While we don't know exactly what exercises Mozart's father used to train his son, we do know that by the time Mozart was six or seven he had trained far more intensely and for far longer than the two dozen children who developed perfect pitch through Sakakibara's practice sessions. In retrospect, then, there should be nothing at all surprising about Mozart's development of perfect pitch.

So did the seven-year-old Wolfgang have a gift for perfect pitch?

Yes and no. Was he born with some rare genetic endowment that allowed him to identify the precise pitch of a piano note or a whistling teakettle? Everything that scientists have learned about perfect pitch says no. Indeed, if Mozart had been raised in some other family without exposure to music — or without enough of the right sort of exposure — he would certainly have never developed that ability at all. Nonetheless, Mozart was indeed born with a gift, and it was the same gift that the children in Sakakibara's study were born with. They were all endowed with a brain so flexible and adaptable that it could, with the right sort of training, develop a capability that seems quite magical to those of us who do not possess it.

In short, perfect pitch is not the gift, but, rather, *the ability to develop perfect pitch* is the gift — and, as nearly as we can tell, pretty much everyone is born with that gift.

This is a wonderful and surprising fact. In the millions of years of evolution leading up to modern humans, there were almost certainly no selection pressures favoring people who could identify, say, the precise notes that a bird was singing. Yet here we are today, able to develop perfect pitch with a relatively simple training regimen.

Only recently have neuroscientists come to understand why such a gift should exist. For decades scientists believed that we were born with our brains' circuits pretty much fixed and that this circuitry determined our abilities. Either your brain was wired for perfect pitch, or it wasn't, and there wasn't much you could do to change it. You might need a certain amount of practice to bring that innate talent into full bloom, and if you didn't get this practice, your perfect pitch might never develop fully, but the general belief was that no amount of practice would help if you didn't have the right genes to start with.

But since the 1990s brain researchers have come to realize that the brain — even the adult brain — is far more adaptable than anyone ever imagined, and this gives us a tremendous amount of control over what our brains are able to do. In particular, the brain responds to the right sorts of triggers by rewiring itself in various ways. New connections are made between neurons, while existing connections can be strengthened or weakened, and in some parts of the brain it is even possible for new neurons to grow. This adaptability explains how the development of perfect pitch was possible in Sakakibara's subjects as well as in Mozart himself: their brains responded to the musical training by developing certain circuits that enabled perfect pitch. We can't yet identify exactly which circuits those are or say what they look like or exactly what they do, but we know they must be there — and we know that they are the product of the training, not of some inborn genetic programming.

In the case of perfect pitch, it seems that the necessary adaptability in the brain disappears by the time a child passes about six years old, so that if the rewiring necessary for perfect pitch has not occurred by then, it will never happen. (Although, as we will see in chapter 8, there are exceptions of a sort, and these exceptions can teach us a great deal about exactly how people take advantage of the brain's adaptability.) This loss is part of a broader phenomenon — that is, that both the brain and the body are more adaptable in young children than in adults, so there are certain abilities that can only be developed, or that are more easily developed, before the age of six or twelve or eighteen. Still, both the brain and the body retain a great deal of adaptability throughout adulthood, and this adaptability makes it possible for adults, even older adults, to develop a wide variety of new capabilities with the right training.

With this truth in mind, let's return to the question that I asked at the beginning: Why are some people so amazingly good at what they do? Over my years of studying experts in various fields, I have found that they all develop their abilities in much the same way that Sakakibara's students did — through dedicated training that drives changes in the brain (and sometimes, depending on the ability, in the body) that make it possible for them to do things that they otherwise could not. Yes, in some cases genetic endowment makes a difference, particularly in areas where height or other physical factors are important. A man with genes for being five feet five will find it tough to become a professional basketball player, just as a six-foot woman will find it virtually impossible to succeed as an artistic gymnast at the international level. And, as we will discuss later in this book, there are other ways in which genes may influence one's achievements, particularly those genes that influence how likely a person is to practice diligently and correctly. But the clear message from decades of research is that no matter what role innate genetic endowment may play in the achievements of "gifted" people, the main gift that these people have is the same one we all have — the adaptability of the human brain and body, which they have taken advantage of more than the rest of us.

If you talk to these extraordinary people, you find that they all understand this at one level or another. They may be unfamiliar with the concept of cognitive adaptability, but they seldom buy into the idea that they have reached the peak of their fields because they were the lucky winners of some genetic lottery. They know what is required to develop the extraordinary skills that they possess because they have experienced it firsthand.

One of my favorite testimonies on this topic came from Ray Allen, a ten-time All-Star in the National Basketball Association and the greatest three-point shooter in the history of that league. Some years back, ESPN columnist Jackie MacMullan wrote an article about Allen as he was approaching his record for most three-point shots made. In talking with Allen for that story, MacMullan mentioned that another basketball commentator had said that Allen was born with a shooting touch — in other words, an innate gift for three-pointers. Allen did not agree.

"I've argued this with a lot of people in my life," he told MacMullan. "When people say God blessed me with a beautiful jump shot, it really pisses me off. I tell those people, 'Don't undermine the work I've put in every day.' Not some days. Every day. Ask anyone who has been on a team with me who shoots the most. Go back to Seattle and Milwaukee, and ask them. The answer is me." And, indeed, as MacMullan noted, if you talk to Allen's high school basketball coach you will find that Allen's jump shot was not noticeably better than his teammates' jump shots back then; in fact, it was poor. But Allen took control, and over time, with hard work and dedication, he transformed his jump shot into one so graceful and natural that people assumed he was born with it. He took advantage of his gift — his real gift.

ABOUT THIS BOOK

This is a book about the gift that Wolfgang Amadeus Mozart, Sakakibara's schoolchildren, and Ray Allen all shared — the ability to create, through the right sort of training and practice, abilities that they would not otherwise possess by taking advantage of the incredible adaptability of the human brain and body. Furthermore, it is a book about how anyone can put this gift to work in order to improve in an area they choose. And finally, in the broadest sense this is a book about a fundamentally new way of thinking about human potential, one that suggests we have far more power than we ever realized to take control of our own lives.

Since antiquity, people have generally assumed that a person's potential in any given field is inevitably and unavoidably limited by that person's inherent talent. Many people take piano lessons, but only those with some special gift become truly great pianists or composers. Every child is exposed to mathematics in school, but only a few have what it takes to become mathematicians or physicists or engineers. According to this view, each of us is born with a set of fixed potentials — a potential for music, a potential for mathematics, a potential for sports, a potential for business — and we can choose to develop (or not) any of those potentials, but we cannot fill any one of those particular "cups" up past its brim. Thus the purpose of teaching or training becomes helping a person reach his or her potential — to fill the cup as fully as possible. This implies a certain approach to learning that assumes preset limits.

But we now understand that there's no such thing as a predefined ability. The brain is adaptable, and training can create skills — such as perfect pitch — that did not exist before. This is a game changer, because learning now becomes a way of creating abilities rather than of bringing people to the point where they can take advantage of their innate ones. In this new world it no longer makes sense to think of people as born with fixed reserves of potential; instead, potential is an expandable vessel, shaped by the various things we do throughout our lives. Learning isn't a way of reaching one's potential but rather a way of developing it. We can create our own potential. And this is true whether our goal is to become a concert pianist or just play the piano well enough to amuse ourselves, to join the PGA golf tour or just bring our handicaps down a few strokes.

The question then becomes, How do we do it? How do we take advantage of this gift and build abilities in our area of choice? Much of my research over the past several decades has been devoted to answering this question — that is, to identify and understand in detail the best ways to improve performance in a given activity. In short, I have been asking, What works and what doesn't and why?

Surprisingly, this question has gotten very little attention from most of the people who have written about this general subject. Over the past few years a number of books have argued that people have been overestimating the value of innate talent and underestimating the value of such things as opportunity, motivation, and effort. I cannot disagree with this, and it is certainly important to let people know that they can improve — and improve a lot — with practice, or else they are unlikely to be motivated to even try. But sometimes these books leave the impression that heartfelt desire and hard work alone will lead to improved performance — "Just keep working at it, and you'll get there"— and this is wrong. The right sort of practice carried out over a sufficient period of time leads to improvement. Nothing else.

This book describes in detail what that "right sort of practice" is and how it can be put to work.

The details about this sort of practice are drawn from a relatively new area of psychology that can be best described as "the science of expertise." This new field seeks to understand the abilities of "expert performers," that is, people who are among the best in the world at what they do, who have reached the very peak of performance, and I have published several academic books on the topic, including *Toward a General Theory of Expertise: Prospects and Limits* in 1991, *The Road to Excellence* in 1996, and *The Cambridge Handbook of Expertise and Expert Performance* in 2006. Those of us in the expertise field investigate what sets these exceptional people apart from everyone else. We also try to assemble a step-by-step accounting of how these expert performers improved their performance over time and exactly how their mental and physical abilities changed as they improved. More than two decades ago, after studying expert performers from a wide range of fields, my colleagues and I came to realize that no matter what the field, the most effective approaches to improving performance all follow a single set of general principles. We named this universal approach "deliberate practice." Today deliberate practice remains the gold standard for anyone in any field who wishes to take advantage of the gift of adaptability in order to build new skills and abilities, and it is the main concern of this book.

The first half of this book describes what deliberate practice is, why it works as well as it does, and how experts apply it to produce their extraordinary abilities. To do that we will have to examine various types of practice, from the least to the most sophisticated, and discuss what differentiates them. Because one of the key differences among different types of practice is the extent to which they harness the adaptability of the human brain and body, we will take some time to discuss that adaptability and what triggers it. We'll also explore exactly what sorts of changes take place in the brain in response to deliberate practice. Because gaining expertise is largely a matter of improving one's mental processes (including, in some fields, the mental processes that control body movements), and because physical changes such as increasing strength, flexibility, and endurance are already reasonably well understood, this book's focus will be mostly on the mental side of expert performance, although there is certainly a significant physical component to expertise in sports and other athletic endeavors. After these explorations we will examine how everything fits together to produce an expert performer — a long-term process that generally takes a decade or more.

Next, in a brief interlude, we examine more closely the issue of innate endowment and what role it might play in limiting how far some people can go in attaining expert performance. There are some inherited physical characteristics, such as height and body size, that can influence performance in various sports and other physical activities and that cannot be changed by practice. However, most traits that play a role in expert performance can be modified by the right sort of practice, at least during some period of one's lifespan. More generally, there is a complex interplay between genetic factors and practice activities that we are just beginning to understand. Some genetic factors may influence a person's ability to engage in sustained deliberate practice — for instance, by limiting a person's capability to focus for long periods of time every day. Conversely, engaging in extended practice may influence how genes are turned on and off in the body.

The last part of the book takes everything we have learned about deliberate practice by studying expert performers and explains what it means for the rest of us. I offer specific advice about putting deliberate practice to work in professional organizations in order to improve the performance of employees, about how individuals can apply deliberate practice to get better in their areas of interest, and even about how schools can put deliberate practice to work in the classroom.

While the principles of deliberate practice were discovered by studying expert performers, the principles themselves can be used by anyone who wants to improve at anything, even if just a little bit. Want to improve your tennis game? Deliberate practice. Your writing? Deliberate practice. Your sales skills? Deliberate practice. Because deliberate practice was developed specifically to help people become among the best in the world at what they do and not merely to become "good enough," it is the most powerful approach to learning that has yet been discovered.

Here is a good way to think about it: You wish to climb a mountain. You're not sure how high you want to go — that peak looks an awfully long way off — but you know you want to get higher than you currently are. You could simply take off on whichever path looks promising and hope for the best, but you're probably not going to get very far. Or you could rely on a guide who has been to the peak and knows the best way there. That will guarantee that no matter how high you decide to climb, you are doing it in the most efficient, effective way. That best way is deliberate practice, and this book is your guide. It will show you the path to the peak; how far you travel along that path is up to you.

PEAK

1

The Power of Purposeful Practice

IN JUST OUR FOURTH SESSION together, Steve was already beginning to sound discouraged. It was Thursday of the first week of an experiment that I had expected to last for two or three months, but from what Steve was telling me, it might not make much sense to go on. "There appears to be a limit for me somewhere around eight or nine digits," he told me, his words captured by the tape recorder that ran throughout each of our sessions. "With nine digits especially, it's very difficult to get regardless of what pattern I use — you know, my own kind of strategies. It really doesn't matter what I use — it seems very difficult to get."

Steve, an undergraduate at Carnegie Mellon University, where I was teaching at the time, had been hired to come in several times a week and work on a simple task: memorizing strings of numbers. I would read him a series of digits at a rate of about one per second — "Seven . . . four . . . zero . . . one . . . one . . . nine . . ." and so on — and Steve would try to remember them all and repeat them back to me once I was done. One goal was simply to see how much Steve could improve with practice. Now, after four of the hour-long sessions, he could reliably recall seven-digit strings — the length of a local phone number — and he usually got the eight-digit strings right, but nine digits was hit or miss, and he had never managed to remember a ten-digit string at all. And at this point, given his frustrating experience over the first few sessions, he was pretty sure that he wasn't going to get any better.

What Steve didn't know — but I did — was that pretty much all of psychological science at the time indicated that he was right. Decades of research had shown that there is a strict limit to the number of items that a person can retain in short-term memory, which is the type of memory the brain uses to hold on to small amounts of information for a brief period of time. If a friend gives you his address, it is your short-term memory that holds on to it just long enough to write it down. Or if you're multiplying a couple of two-digit numbers in your head, your short-term memory is where you keep track of all the intermediate pieces: "Let's see: 14 times 27 . . . First, 4 times 7 is 28, so keep the 8 and carry the 2, then 4 times 2 is 8 . . ." and so on. And there's a reason it's called "short-term." You're not going to remember that address or those intermediate numbers five minutes later unless you spend the time repeating them to yourself over and over again — and thus transfer them into your long-term memory.

The problem with short-term memory — and the problem that Steve was coming face-to-face with — is that the brain has strict limits on how many items it can hold in short-term memory at once. For some it is six items, for others it may be seven or eight, but the limit is generally about seven items — enough to hold on to a local phone number but not a Social Security number. Long-term memory doesn't have the same limitations — in fact, no one has ever found the upper limits of long-term memory — but it takes much longer to deploy. Given enough time to work on it, you can memorize dozens or even hundreds of phone numbers, but the test I was giving Steve was designed to present digits so fast that he was forced to use only his short-term memory. I was reading the digits at a rate of one per second — too fast for him to transfer the digits into his long-term memory — so it was no surprise that he was running into a wall at numbers that were about eight or nine digits long.

Still, I hoped he might be able to do a little better. The idea for the study had come from an obscure paper I had discovered while searching through old scientific studies, a paper published in a 1929 issue of the *American Journal of Psychology* by Pauline Martin and Samuel Fernberger, two psychologists at the University of Pennsylvania. Martin and Fernberger reported that two undergraduate subjects had been able, with four months of practice, to increase the number of digits they could remember when given the digits at a rate of about one per second. One of the students had improved from an average of nine digits to thirteen, while the other had gone from eleven to fifteen.

This result had been overlooked or forgotten by the broader psychological research community, but it immediately captured my attention. Was this sort of improvement really even possible? And, if so, *how* was it possible? Martin and Fernberger had offered no details about how the students had improved their digit memory, but that was exactly the sort of question that most intrigued me. At the time, I was just out of graduate school, and my main area of interest was the mental processes that take place when someone is learning something or developing a skill. For my dissertation I had honed a psychological research tool called "the think-aloud protocol" that was designed specifically to study such mental processes. So in collaboration with Bill Chase, a well-known Carnegie Mellon psychology professor, I set out to redo the old Martin and Fernberger study, and this time I would be watching to see exactly how our subject improved his digit memory — if indeed he did.

The subject we had recruited was Steve Faloon, who was about as typical a Carnegie Mellon undergraduate as we could have hoped to find. He was a psychology major who was interested in early childhood development. He had just finished his junior year. His scores on achievement tests were similar to those of other Carnegie Mellon students, while his grades were somewhat higher than average. Tall and thin with thick, dark-blond hair, he was friendly, outgoing, and enthusiastic. And he was a serious runner — a fact that did not seem meaningful to us at the time but that would turn out to be crucial to our study.

On the first day that Steve showed up for the memory work, his performance was dead-on average. He could usually remember seven digits and sometimes eight but no more. It was the same sort of performance you would expect from any random person picked off the street. On Tuesday, Wednesday, and Thursday he was a little better — an average of just under nine digits — but still no better than normal. Steve said he thought that the main difference from the first day was that he knew what to expect from the memory test and thus was more comfortable. It was at the end of that Thursday's session that Steve explained to me why he thought he was unlikely to get any better.

Then on Friday something happened that would change everything. Steve found a way to break through. The training sessions went like this: I would start with a random five-digit string, and if Steve got it right (which he always did), I would go to six digits. If he got that right, we'd go to seven digits, and so on, increasing the length of the string by one each time he got it right. If he got it wrong, I would drop the length of the string by two and go again. In this way Steve was constantly challenged, but not too much. He was given strings of digits that were right at the boundary between what he could and couldn't do.

And on that Friday, Steve moved the boundary. Up to that point he had remembered a nine-digit string correctly only a handful of times, and he had never remembered a ten-digit string correctly, so he had never even had a chance to try strings of eleven digits or longer. But he began that fifth session on a roll. He got the first three tries — five, six, and seven digits — right without a problem, missed the fourth one, then got back on track: six digits, right; seven digits, right; eight digits, right; nine digits, right. Then I read out a ten-digit number — 5718866610 — and he nailed that one as well. He missed the next string with eleven digits, but after he got another nine digits and another ten digits right, I read him a second eleven-digit string — 90756629867 — and this time he repeated the whole thing back to me without a hitch. It was two digits more than he had ever gotten right before, and although an additional two digits may not seem particularly impressive, it was actually a major accomplishment because the past several days had established that Steve had a "natural" ceiling — the number of digits he could comfortably hold in his short-term memory — of only eight or nine. He had found a way to push through that ceiling.

That was the beginning of what was to be the most surprising two years of my career. From this point on, Steve slowly but steadily improved his ability to remember strings of digits. By the sixtieth session he was able to consistently remember twenty digits — far more than Bill and I had imagined he ever could. After a little more than one hundred sessions, he was up to forty, which was more than anyone, even professional mnemonists, had ever achieved, and still he kept going. He worked with me for more than two hundred training sessions, and by the end he had reached eighty-two digits — eighty-two! If you think about that for a moment, you'll realize just how incredible this memory ability truly is. Here are eighty-two random digits:

0326443449602221328209301020391832373927788917267653245037746120179094345510355530

Imagine hearing all of those read out to you at one per second and *being able to remember them all.* This is what Steve Faloon taught himself to do over the two years of our experiment — all without even knowing it was possible, just by continuing to work on it week after week.

THE RISE OF EXTRAORDINARY PERFORMERS

In 1908 Johnny Hayes won the Olympic marathon in what newspapers at the time described as "the greatest race of the century." Hayes's winning time, which set a world record for the marathon, was 2 hours, 55 minutes, and 18 seconds.

Today, barely more than a century later, the world record for a marathon is 2 hours, 2 minutes, and 57 seconds — nearly 30 percent faster than Hayes's record time — and if you're an eighteen- to thirty-four-year-old male, you aren't even allowed to enter the Boston Marathon unless you've run another marathon in less than 3 hours, 5 minutes. In short, Hayes's world-record time in 1908 would qualify him for today's Boston Marathon (which has about thirty thousand runners) but with not a lot to spare.

That same 1908 Summer Olympics saw a near disaster in the men's diving competition. One of the divers barely avoided serious injury while attempting a double somersault, and an official report released a few months later concluded that the dive was simply too dangerous and recommended that it be banned from future Olympic Games. Today the double somersault is an entry-level dive, with ten-year-olds nailing it in competitions, and by high school the best divers are doing four and a half somersaults. World-class competitors take it even further with dives such as "the Twister"— two and a half backward somersaults with two and a half twists added. It's difficult to imagine what those early-twentieth-century experts who found the double-somersault dive too dangerous would have thought about the Twister, but my guess is that they would have dismissed it as laughably impossible — assuming, that is, that someone would have had the imagination and the audacity to suggest it in the first place.

In the early 1930s Alfred Cortot was one of the best-known classical musicians in the world, and his recordings of Chopin's "24 Études" were considered the definitive interpretation. Today teachers offer those same performances — sloppy and marred by missed notes — as an example of how *not* to play Chopin, with critics complaining about Cortot's careless technique, and any professional pianist is expected to be able to perform the études with far greater technical skill and élan than Cortot. Indeed, Anthony Tommasini, the music critic at the *New York Times,* once commented that musical ability has increased so much since Cortot's time that Cortot would probably not be admitted to Juilliard now.

In 1973 David Richard Spencer of Canada had memorized more digits of pi than any person before him: 511. Five years later, after a rapid-fire series of new records set by a handful of people competing to claim the memorization title, the record belonged to an American, David Sanker, who had committed 10,000 digits of pi to memory. In 2015, after another thirty-plus years of gains, the recognized title holder was Rajveer Meena of India, who had memorized the first 70,000 digits of pi — an accumulation that took him 24 hours and 4 minutes to recite — although Akira Haraguchi of Japan had claimed to have memorized an even more incredible 100,000 digits, or nearly two hundred times as many as anyone had memorized just forty-two years earlier.

These are not isolated examples. We live in a world full of people with extraordinary abilities — abilities that from the vantage point of almost any other time in human history would have been deemed impossible. Consider Roger Federer's magic with a tennis ball, or the astounding vault that McKayla Maroney nailed in the 2012 Summer Olympics: a round-off onto the springboard, a back handspring onto the vault, and then a high, arching flight with McKayla completing two and a half twists before she landed firmly and with complete control on the mat. There are chess grandmasters who can play several dozen different games simultaneously — while blindfolded — and a seemingly unending supply of young musical prodigies who can do things on the piano, the violin, the cello, or the flute that would have astonished aficionados a century ago.

But while the abilities are extraordinary, there is no mystery at all about how these people developed them. They practiced. A lot. The world-record time in the marathon wasn't cut by 30 percent over the course of a century because people were being born with a greater talent for running long distances. Nor did the second half of the twentieth century see some sudden surge in the births of people with a gift for playing Chopin or Rachmaninoff or for memorizing tens of thousands of random digits.

What the second half of the twentieth century did see was a steady increase in the amount of time that people in different areas devoted to training, combined with a growing sophistication of training techniques. This was true in a huge number of fields, particularly highly competitive fields such as musical performance and dance, individual and team sports, and chess and other competitive games. This increase in the amount and sophistication of practice resulted in a steady improvement in the abilities of the performers in these various fields — an improvement that was not always obvious from year to year but that is dramatic when viewed over the course of several decades.

One of the best, if sometimes bizarre, places to see the results of this sort of practice is in *Guinness World Records.* Flip through the pages of the book or visit the online version, and you will find such record holders as the American teacher Barbara Blackburn, who can type up to 212 words per minute; Marko Baloh of Slovenia, who once rode 562 miles on a bicycle in twenty-four hours; and Vikas Sharma of India, who in just one minute was able to calculate the roots of twelve large numbers, each with between twenty and fifty-one digits, with the roots ranging from the seventeenth to the fiftieth root. That last may be the most impressive of all of them because Sharma was able to perform twelve exceedingly difficult mental calculations in just sixty seconds — faster than many people could punch the numbers into a calculator and read off the answers.

I actually received an e-mail from one Guinness world record holder, Bob J. Fisher, who at one time held twelve different world records for basketball free-throw shooting. His records include such things as the most free throws accomplished in thirty seconds (33), the most in ten minutes (448), and the most in one hour (2,371). Bob wrote to tell me that he had read about my studies of the effects of practice and had applied what he had learned from those studies in developing his ability to shoot basketball free throws faster than anyone else.

Those studies all have their roots in the work that I did with Steve Faloon in the late 1970s. Since that time I have devoted my career to understanding exactly how practice works to create new and expanded capabilities, with a particular focus on those people who have used practice to become among the best in the world at what they do. And after several decades of studying these best of the best — these "expert performers," to use the technical term — I have found that no matter what field you study, music or sports or chess or something else, the most effective types of practice all follow the same set of general principles.

There is no obvious reason why this should be the case. Why should the teaching techniques used to turn aspiring musicians into concert pianists have anything to do with the training that a dancer must go through to become a prima ballerina or the study that a chess player must undertake to become a grandmaster? The answer is that the most effective and most powerful types of practice in any field work by harnessing the adaptability of the human body and brain to create, step by step, the ability to do things that were previously not possible. If you wish to develop a truly effective training method for anything — creating world-class gymnasts, for instance, or even something like teaching doctors to perform laparoscopic surgery — that method will need to take into account what works and what doesn't in driving changes in the body and brain. Thus, all truly effective practice techniques work in essentially the same way.

These insights are all relatively new and weren't available to all the teachers, coaches, and performers who produced the incredible improvements in performance that have occurred over the past century. Instead, these advances were all accomplished through trial and error, with the people involved having essentially no idea why a particular training method might be effective. Furthermore, the practitioners in the various fields built their bodies of knowledge in isolation, with no sense that all of this was interconnected — that the ice-skater who was working on a triple axel was following the same set of general principles as, say, the pianist working to perfect a Mozart sonata. So imagine what might be possible with efforts that are inspired and directed by a clear scientific understanding of the best ways to build expertise. And imagine what might be possible if we applied the techniques that have proved to be so effective in sports and music and chess to all the different types of learning that people do, from the education of schoolchildren to the training of doctors, engineers, pilots, businesspeople, and workers of every sort. I believe that the dramatic improvements we have seen in those few fields over the past hundred years are achievable in pretty much every field if we apply the lessons that can be learned from studying the principles of effective practice.

There are various sorts of practice that can be effective to one degree or another, but one particular form — which I named "deliberate practice" back in the early 1990s — is the gold standard. It is the most effective and powerful form of practice that we know of, and applying the principles of deliberate practice is the best way to design practice methods in any area. We will devote most of the rest of this book to exploring what deliberate practice is, why it is so effective, and how best to apply it in various situations. But before we delve into the details of deliberate practice, it will be best if we spend a little time understanding some more basic types of practice — the sorts of practice that most people have already experienced in one way or another.

THE USUAL APPROACH

Let's begin by looking at the way people typically learn a new skill — driving a car, playing the piano, performing long division, drawing a human figure, writing code, or pretty much anything, really. For the sake of having a specific example, let's suppose you are learning to play tennis.

You've seen tennis matches played on television, and it looks like fun, or maybe you have some friends who play tennis and want you to join them. So you buy a couple of tennis outfits, court shoes, maybe a sweatband, and a racket and some balls. Now you're committed, but you don't know the first thing about actually playing tennis — you don't even know how to hold the racket — so you pay for some lessons from a tennis coach or maybe you just ask one of your friends to show you the basics. After those initial lessons you know enough to go out on your own and practice. You'll probably spend some time working on your serve, and you practice hitting the ball against a wall over and over again until you're pretty sure you can hold your own in a game against a wall. After that you go back to your coach or your friend for another lesson, and then you practice some more, and then another lesson, more practice, and after a while you've reached the point where you feel competent enough to play against other people. You're still not very good, but your friends are patient, and everyone has a good time. You keep practicing on your own and getting a lesson every now and then, and over time the really embarrassing mistakes — like swinging and missing the ball completely or hitting the ball very solidly straight into your doubles partner's back — become more and more rare. You get better with the various strokes, even the backhand, and occasionally, when everything comes together just so, you even end up hitting the ball like a pro (or so you tell yourself). You have reached a comfort level at which you can just go out and have fun playing the game. You pretty much know what you're doing, and the strokes have become automatic. You don't have to think too much about any of it. So you play weekend after weekend with your friends, enjoying the game and the exercise. You have become a tennis player. That is, you have "learned" tennis in the traditional sense, where the goal is to reach a point at which everything becomes automatic and an acceptable performance is possible with relatively little thought, so that you can just relax and enjoy the game.

At this point, even if you're not completely satisfied with your level of play, your improvement stalls. You have mastered the easy stuff.

But, as you quickly discover, you still have weaknesses that don't disappear no matter how often you play with your friends. Perhaps, for example, every time you use a backstroke to hit a ball that is coming in chest-high with a bit of spin, you miss the shot. You know this, and the cagier of your opponents have noticed this too, so it is frustrating. However, because it doesn't happen very often and you never know when it's coming, you never get a chance to consciously work on it, so you keep missing the shot in exactly the same way as you manage to hit other shots — automatically.

We all follow pretty much the same pattern with any skill we learn, from baking a pie to writing a descriptive paragraph. We start off with a general idea of what we want to do, get some instruction from a teacher or a coach or a book or a website, practice until we reach an acceptable level, and then let it become automatic. And there's nothing wrong with that. For much of what we do in life, it's perfectly fine to reach a middling level of performance and just leave it like that. If all you want to do is to safely drive your car from point A to point B or to play the piano well enough to plink out "Für Elise," then this approach to learning is all you need.

But there is one very important thing to understand here: once you have reached this satisfactory skill level and automated your performance — your driving, your tennis playing, your baking of pies — you have stopped improving. People often misunderstand this because they assume that the continued driving or tennis playing or pie baking is a form of practice and that if they keep doing it they are bound to get better at it, slowly perhaps, but better nonetheless. They assume that someone who has been driving for twenty years must be a better driver than someone who has been driving for five, that a doctor who has been practicing medicine for twenty years must be a better doctor than one who has been practicing for five, that a teacher who has been teaching for twenty years must be better than one who has been teaching for five.

But no. Research has shown that, generally speaking, once a person reaches that level of "acceptable" performance and automaticity, the additional years of "practice" don't lead to improvement. If anything, the doctor or the teacher or the driver who's been at it for twenty years is likely to be a bit worse than the one who's been doing it for only five, and the reason is that these automated abilities gradually deteriorate in the absence of deliberate efforts to improve.

So what do you do if you're not satisfied with this automated level of performance? What if you are a teacher with ten years in the classroom and you want to do something to better engage your students and get your lessons across more effectively? A weekend golfer and you would like to move beyond your eighteen handicap? An advertising copywriter and you want to add a little wow to your words?

This is the same situation that Steve Faloon found himself in after just a couple of sessions. At that point he had become comfortable with the task of hearing a string of digits, holding them in his memory, and repeating them back to me, and he was performing about as well as could be expected, given what is known about the limitations of short-term memory. He could have just kept doing what he was doing and maxing out at eight or nine digits, session after session. But he didn't, because he was participating in an experiment in which he was constantly being challenged to remember just one more digit than the last time, and because he was naturally the sort of guy who liked this sort of challenge, Steve pushed himself to get better.

The approach that he took, which we will call "purposeful practice," turned out to be incredibly successful for him. It isn't always so successful, as we shall see, but it is more effective than the usual just-enough method — and it is a step toward deliberate practice, which is our ultimate goal.

PURPOSEFUL PRACTICE

Purposeful practice has several characteristics that set it apart from what we might call "naive practice," which is essentially just doing something repeatedly, and expecting that the repetition alone will improve one's performance.

Steve Oare, a specialist in music education at Wichita State University, once offered the following imaginary conversation between a music instructor and a young music student. It's the sort of conversation about practice that music instructors have all the time. In this case a teacher is trying to figure out why a young student has not been improving:

TEACHER: Your practice sheet says that you practice an hour a day, but your playing test was only a C. Can you explain why?
STUDENT: I don't know what happened! I could play the test last night!
TEACHER: How many times did you play it?
STUDENT: Ten or twenty.
TEACHER: How many times did you play it correctly?
STUDENT: Umm, I dunno . . . Once or twice . . .
TEACHER: Hmm . . . How did you practice it?
STUDENT: I dunno. I just played it.

This is naive practice in a nutshell: I just played it. I just swung the bat and tried to hit the ball. I just listened to the numbers and tried to remember them. I just read the math problems and tried to solve them.

Purposeful practice is, as the term implies, much more purposeful, thoughtful, and focused than this sort of naive practice. In particular, it has the following characteristics:

Purposeful practice has well-defined, specific goals. Our hypothetical music student would have been much more successful with a practice goal something like this: "Play the piece all the way through at the proper speed without a mistake three times in a row." Without such a goal, there was no way to judge whether the practice session had been a success.

In Steve's case there was no long-range goal because none of us knew how many digits one could possibly memorize, but he had a very specific short-term goal: to remember more digits than he had the previous session. As a distance runner, Steve was very competitive, even if he was only competing with himself, and he brought that attitude to the experiment. From the very beginning Steve was pushing each day to increase the number of digits he could remember.

Purposeful practice is all about putting a bunch of baby steps together to reach a longer-term goal. If you're a weekend golfer and you want to decrease your handicap by five strokes, that's fine for an overall purpose, but it is not a well-defined, specific goal that can be used effectively for your practice. Break it down and make a plan: What exactly do you need to do to slice five strokes off your handicap? One goal might be to increase the number of drives landing in the fairway. That's a reasonably specific goal, but you need to break it down even more: What exactly will you do to increase the number of successful drives? You will need to figure out why so many of your drives are not landing in the fairway and address that by, for instance, working to reduce your tendency to hook the ball. How do you do that? An instructor can give you advice on how to change your swing motion in specific ways. And so on. The key thing is to take that general goal — get better — and turn it into something specific that you can work on with a realistic expectation of improvement.

Purposeful practice is focused. Unlike the music student that Oare described, Steve Faloon was focused on his task from the very beginning, and his focus grew as the experiment went along and he was memorizing longer and longer strings of digits. You can get a sense of this focus by listening to the tape of session 115, which came about halfway through the study. Steve had regularly been remembering strings of close to forty digits, but forty itself was not something he could yet do with any consistency, and he really wanted to reach forty regularly on this day. We began with thirty-five digits, which was easy for him, and he started pumping himself up as the strings increased in length. Before I read the thirty-nine-digit string, he gave himself an excited pep talk, seemingly conscious of nothing but the approaching task: "We have a big day here! . . . I haven't missed one yet, have I? No! . . . This will be a banner day!" He was silent during the forty seconds it took me to read out the numbers, but then, as he carefully went over the digits in his head, remembering various groups of them and the order in which they appeared, he could barely contain himself. He hit the table loudly a number of times, and he clapped a lot, apparently in celebration of remembering this or that group of digits or where they went in the string. Once he blurted out, "Absolutely right! I'm certain!" And when he finally spit the digits back at me, he was indeed right, so we moved on to forty. Again, the pep talk: "Now this is the big one! If I get past this one, it's all over! I have to get past this one!" Again the silence as I read the digits, and then the excited noises and exclamations as he cogitated. "Wow! . . . Come on now! . . . All right! . . . Go!" He got that one right as well, and the session indeed became one in which he regularly hit forty digits, although no more.

Now, not everyone will focus by hollering and pounding on a table, but Steve's performance illustrates a key insight from the study of effective practice: You seldom improve much without giving the task your full attention.

Purposeful practice involves feedback. You have to know whether you are doing something right and, if not, how you're going wrong. In Oare's example the music student got belated feedback at school with a C on the performance test, but there seems to have been no feedback during practice — no one listening and pointing out mistakes, with the student seemingly clueless about whether there were errors in the practice. ("How many times did you play it correctly?" "Umm, I dunno . . . Once or twice . . .")

In our memory study, Steve got simple, direct feedback after every attempt — correct or incorrect, success or failure. He always knew where he stood. But perhaps the more important feedback was something that he did himself. He paid close attention to which aspects of a string of digits caused him problems. If he'd gotten the string wrong, he usually knew exactly why and which digits he had messed up on. Even if he got the string correct, he could report to me afterward which digits had given him trouble and which had been no problem. By recognizing where his weaknesses were, he could switch his focus appropriately and come up with new memorization techniques that would address those weaknesses.

Generally speaking, no matter what you're trying to do, you need feedback to identify exactly where and how you are falling short. Without feedback — either from yourself or from outside observers — you cannot figure out what you need to improve on or how close you are to achieving your goals.

Purposeful practice requires getting out of one's comfort zone. This is perhaps the most important part of purposeful practice. Oare's music student shows no sign of ever pushing himself beyond what was familiar and comfortable. Instead, the student's words seem to imply a rather desultory attempt at practice, with no effort to do more than what was already easy for him. That approach just doesn't work.

Our memory experiment was set up to keep Steve from getting too comfortable. As he increased his memory capacity, I would challenge him with longer and longer strings of digits so that he was always close to his capacity. In particular, by increasing the number of digits each time he got a string right, and decreasing the number when he got it wrong, I kept the number of digits right around what he was capable of doing while always pushing him to remember just one more digit.

This is a fundamental truth about any sort of practice: If you never push yourself beyond your comfort zone, you will never improve. The amateur pianist who took half a dozen years of lessons when he was a teenager but who for the past thirty years has been playing the same set of songs in exactly the same way over and over again may have accumulated ten thousand hours of "practice" during that time, but he is no better at playing the piano than he was thirty years ago. Indeed, he's probably gotten worse.

We have especially strong evidence of this phenomenon as it applies to physicians. Research on many specialties shows that doctors who have been in practice for twenty or thirty years do worse on certain objective measures of performance than those who are just two or three years out of medical school. It turns out that most of what doctors do in their day-to-day practice does nothing to improve or even maintain their abilities; little of it challenges them or pushes them out of their comfort zones. For that reason, I participated in a consensus conference in 2015 to identify new types of continuing medical education that will challenge doctors and help them maintain and improve their skills. We will discuss this in detail in chapter 5.

Perhaps my favorite example of this lesson is the case of Ben Franklin's chess skills. Franklin was America's first famous genius. He was a scientist who made his reputation with his studies of electricity, a popular writer and publisher of *Poor Richard's Almanack,* the founder of the first public lending library in America, an accomplished diplomat, and the inventor of, among other things, bifocals, the lightning rod, and the Franklin stove. But his greatest passion was chess. He was one of the first chess players in America, and he was a participant in the earliest game of chess known to have been played here. He played chess for more than fifty years, and as he got older he spent more and more time on it. While in Europe he played with François-André Danican Philidor, the best chess player of the time. And despite his well-known advice to be early to bed and early to rise, Franklin regularly played from around 6:00 p.m. until sunrise.

So Ben Franklin was brilliant, and he spent thousands of hours playing chess, sometimes against the best players of the time. Did that make him a great chess player? No. He was above average, but he never got good enough to compare with Europe's better players, much less the best. This failing was a source of great frustration to him, but he had no idea why he couldn't get any better. Today we understand: he never pushed himself, never got out of his comfort zone, never put in the hours of purposeful practice it would take to improve. He was like the pianist playing the same songs the same way for thirty years. That is a recipe for stagnation, not improvement.

Getting out of your comfort zone means trying to do something that you couldn't do before. Sometimes you may find it relatively easy to accomplish that new thing, and then you keep pushing on. But sometimes you run into something that stops you cold and it seems like you'll never be able to do it. Finding ways around these barriers is one of the hidden keys to purposeful practice.

Generally the solution is not "try harder" but rather "try differently." It is a technique issue, in other words. In Steve's case, one barrier came when he hit twenty-two digits. He was grouping them into four four-digit groups, which he used various mnemonic tricks to remember, plus a six-digit rehearsal group at the end that he would repeat over and over to himself until he could remember it by the sound of the numbers. But he couldn't figure out how to get past twenty-two digits, because when he tried to hold five four-digit groups in his head, he became confused about their order. He eventually hit upon the idea of using both three-digit groups and four-digit groups, a breakthrough that eventually allowed him to work up to using four four-digit groups, four three-digit groups, and a six-digit rehearsal group, for a maximum of thirty-four digits. Then, once he reached that limit, he had to develop another technique. This was a regular pattern throughout the entire memory study: Steve would improve up to a point, get stuck, look around for a different approach that could help him get past the barrier, find it, and then improve steadily until another barrier arose.

The best way to get past any barrier is to come at it from a different direction, which is one reason it is useful to work with a teacher or coach. Someone who is already familiar with the sorts of obstacles you're likely to encounter can suggest ways to overcome them.

And sometimes it turns out that a barrier is more psychological than anything else. The famous violin teacher Dorothy DeLay once described the time that one of her students came to her to help increase his speed on a particular piece that he was scheduled to play at a music festival. He could not play it fast enough, he told her. How fast, she asked, would you like to play it? He answered that he wanted to play it as fast as Itzhak Perlman, the world-famous violinist. So DeLay first got a recording of Perlman playing the piece and timed it. Then she set a metronome to a slow speed and had her student play the piece at that pace, which was well within his abilities. She had him play it again and again, each time speeding up the metronome a bit. And each time he nailed it. Finally, after he had gone through the piece flawlessly once more, she showed him the setting on the metronome: He had actually played it faster than Perlman.

Bill Chase and I used a similar technique with Steve a couple of times when he had hit a barrier and thought he might not be able to improve further. Once, I slowed down the rate at which I read the digits just a bit, and the extra time made it possible for Steve to remember significantly more digits. This convinced him that the problem was not the number of digits but rather how quickly he was encoding the digits — that is, coming up with mnemonics for the various groups of digits that made up the entire string — and that he could improve his performance if he could just speed up the time he took to commit the digits to long-term memory.

Another time, I gave Steve strings that were ten digits longer than any of the ones he had managed to remember up to that point.

He surprised himself by remembering most of the digits in those strings — and, in particular, remembering more total digits than he had ever done before, even though he wasn't perfect. This convinced him that it was indeed possible to remember longer strings of digits. He realized his problem was not that he had reached the limit of his memory, but rather that he was messing up on one or two groups of digits in the entire string. He decided that the key to moving on was to encode the small groups of digits more carefully, and he began improving again.

Whenever you're trying to improve at something, you will run into such obstacles — points at which it seems impossible to progress, or at least where you have no idea what you should do in order to improve. This is natural. What is not natural is a true dead-stop obstacle, one that is impossible to get around, over, or through. In all of my years of research, I have found it is surprisingly rare to get clear evidence in any field that a person has reached some immutable limit on performance. Instead, I've found that people more often just give up and stop trying to improve.

One caveat here is that while it is always possible to keep going and keep improving, it is not always easy. Maintaining the focus and the effort required by purposeful practice is hard work, and it is generally not fun. So the issue of motivation inevitably comes up: Why do some people engage in this sort of practice? What keeps them going? We will return to these vital questions again and again throughout the book.

In Steve's case, there were several factors at work. First, he was getting paid. But he could have always shown up for the sessions and not tried particularly hard and still have gotten paid, so while that may have been part of his motivation, it was certainly not all of it. Why did he push himself so hard to improve? From talking to him, I believe that a large part of it was that once he started to see improvement after the first few sessions, he really enjoyed seeing his memory scores go up. It felt good, and he wanted to keep feeling that way. Also, after he reached a certain level in his memorization abilities, he became something of a celebrity; stories about him appeared in newspapers and magazines, and he made a number of appearances on television, including the *Today* show. This provided another type of positive feedback. Generally speaking, meaningful positive feedback is one of the crucial factors in maintaining motivation. It can be internal feedback, such as the satisfaction of seeing yourself improve at something, or external feedback provided by others, but it makes a huge difference in whether a person will be able to maintain the consistent effort necessary to improve through purposeful practice.

One other factor was that Steve liked to challenge himself. This was clear from his record as a cross-country and track runner. Everyone who knew him would tell you that he trained as hard as anyone but that his motivation was simply to improve his own performance, not necessarily to win races. Furthermore, from years of running he knew what it meant to train regularly, week after week, month after month, and it seems unlikely that the task of working on his memory three times a week for an hour each time seemed particularly daunting, given that he regularly went for three-hour runs. Later, after finishing the memory work with Steve and a couple of other students, I made it a point to recruit only subjects who had trained extensively as athletes, dancers, musicians, or singers. None of them ever quit on me.

So here we have purposeful practice in a nutshell: Get outside your comfort zone but do it in a focused way, with clear goals, a plan for reaching those goals, and a way to monitor your progress. Oh, and figure out a way to maintain your motivation.

This recipe is an excellent start for anyone who wishes to improve — but it is still just a start.

THE LIMITS OF PURPOSEFUL PRACTICE

While Bill Chase and I were still carrying out our two-year memory study with Steve Faloon — but after Steve had begun to set records with his digit-span memory — we decided to look for another subject who would be willing to take on the same challenge. Neither of us believed that Steve had been born with some special gift for memorizing digits, but rather we assumed that the skills he developed could be attributed completely to the training that he went through, and the best way to prove that was to run the same study with another subject and see if we got the same result.

The first person to volunteer was a graduate student, Renée Elio. Before getting started she was told that her predecessor had dramatically increased the number of digits he could memorize, so she knew such improvement was possible — which was more than Steve had known when he started — but we told Renée nothing about how Steve had done it. She would have to come up with her own approach.

When she started out, she improved at a pace that was very similar to Steve's, and she was able to increase her digit-span memory to close to twenty digits after about fifty hours of practice sessions. However, unlike Steve, at this point she hit a wall that she just couldn't get past. After spending another fifty hours or so without improving, she decided to drop out of the training sessions. She had increased her memory for digits to the point that it was far better than any untrained person — and comparable with some mnemonists — but she fell far short of what Steve had accomplished.

What was the difference? Steve had succeeded by developing a collection of mental structures — various mnemonics, many of them based on running times, plus a system for keeping track of the order of the mnemonics — that allowed him to use his long-term memory to sidestep the usual limitations of short-term memory and remember long strings of digits. When he heard the digits 907, for instance, he conceptualized them as a pretty good two-mile time — 9:07, or 9 minutes, 7 seconds — and they were no longer random numbers that he had to commit to short-term memory but rather something he was already familiar with. As we shall see, the key to improved mental performance of almost any sort is the development of mental structures that make it possible to avoid the limitations of short-term memory and deal effectively with large amounts of information at once. Steve had done this.

Renée, not knowing how Steve had done it, had developed a completely different approach to memorizing the digits. Where Steve had memorized groups of three and four digits mainly in terms of running times, Renée employed an elaborate set of mnemonics that relied on such things as days, dates, and times of day. One key difference between Steve and Renée was that Steve had always decided ahead of time what pattern he would use in memorizing the digits, breaking the strings into three- and four-digit sets plus a group at the end with four to six digits that he would repeat to himself over and over until he had the sound of it in his memory. For twenty-seven digits, for instance, he would organize the digits into three sets of four digits each, three sets of three digits each, and then a six-digit group at the end. We referred to this pre-fixed pattern as a "retrieval structure," and it allowed Steve to focus on memorizing the three- and four-digit sets individually and then keep in mind where in the retrieval structure each of these individual sets fit. This proved to be a very powerful approach, as it allowed him to encode each set of three or four digits as a running time or some other mnemonic, put it in his long-term memory, and then not have to think about it again until he went back at the end to recall all of the digits in the string.

Renée, by contrast, devised her mnemonics on the fly, deciding according to the digits she heard what mnemonic she would use to remember them. For a string like 4778245 she might remember it as April 7, 1978 at 2:45, but if the string was 4778295, she would have to use April 7, 1978 and then start a new date: February 9 . . . Without the sort of consistency that Steve's approach offered, she could not master more than twenty digits.

After that experience Bill and I decided to look for another subject who would be as similar to Steve as possible in terms of the way he would memorize the digit strings. Thus we recruited another runner, Dario Donatelli, a member of the Carnegie Mellon long-distance team and one of Steve's training partners. Steve had told Dario that we were looking for someone who would commit to being a long-term participant in our memory-training study, and Dario agreed.

This time, instead of letting Dario figure it out for himself, we had Steve teach Dario his method for encoding digits. With this head start, Dario was able to improve much more quickly than Steve had, at least initially. He got to twenty digits in significantly fewer training sessions, but he began to slow down after that, and once he reached thirty digits it seemed that he was no longer getting much benefit from following Steve's method, and his progress languished. At that point Dario began developing his own version of Steve's method. He came up with slightly different ways of encoding the strings of three and four digits, and, more importantly, he designed a significantly different retrieval structure that worked much better for him. Still, when we tested how Dario was memorizing the digits, we found that he was relying on mental processes that were very much like the ones that Steve had developed, using long-term memory to sidestep the limitations of short-term memory. After several years of training, Dario would eventually be able to remember more than one hundred digits, or about twenty more than Steve. At this point Dario had become, like Steve before him, the best at this particular skill that the world had ever known.

There is an important lesson here: Although it is generally possible to improve to a certain degree with focused practice and staying out of your comfort zone, that's not all there is to it. Trying hard isn't enough. Pushing yourself to your limits isn't enough. There are other, equally important aspects to practice and training that are often overlooked.

One particular approach to practice and training has proven to be the most powerful and effective way to improve one's abilities in every area that has been studied. This approach is deliberate practice, and we will describe it in detail shortly. But first we'll take a closer look at what is behind the amazing sorts of improvement that are possible with the right sort of practice.

2

Harnessing Adaptability

IF YOU'RE A BODYBUILDER or just someone lifting weights to add some muscle, it is easy to track the results as you challenge your biceps, triceps, quadriceps, pecs, delts, lats, traps, abs, glutes, calves, and hamstrings. A tape measure works, or you can simply look in the mirror and admire your progress. If you're running or biking or swimming to increase your endurance, you can track your progress by your heart rate, your breathing, and how long you can keep going until your muscles falter due to lactic acid buildup.

But if your challenge is mental — becoming proficient at calculus, say, or learning how to play a musical instrument or speak a new language — it's different. There is no easy way to observe the resulting changes in your brain as it adapts to the increasing demands being placed on it. There is no soreness in your cortex the day after a particularly tough training session. You don't have to go out and buy new hats because the old ones are now too small. You don't develop a six-pack on your forehead. And because you can't see any changes in your brain, it's easy to assume that there really isn't much going on.

That would be a mistake, however. There is a growing body of evidence that both the structure and the function of the brain change in response to various sorts of mental training, in much the same way as your muscles and cardiovascular system respond to physical training. With the help of such brain-imaging techniques as magnetic resonance imaging (MRI), neuroscientists have begun to study how the brains of people with particular skills differ from the brains of people without those skills and to explore which sorts of training produce which types of changes. Although there is still a tremendous amount to learn in this area, we already know enough to have a clear idea of how purposeful practice and deliberate practice work to increase both our physical and mental capabilities and make it possible to do things that we never could before.

Much of what we know about how the body adapts to training comes from studies of runners, weightlifters, and various other athletes. Interestingly enough, however, some of the best studies to date of how the brain changes in response to extended training were carried out not with musicians or chess players or mathematicians — some of the more traditional subjects in studies of the effects of practice on performance — but instead with taxi drivers.

THE BRAINS OF LONDON CABBIES

Few cities in the world can baffle a GPS system like London can. To start with, there is no grid of thoroughfares that can be used for orientation and routing as you will find in Manhattan or Paris or Tokyo. Instead the city's major streets are set at odd angles to each other. They curve and they squiggle. One-way streets abound, there are traffic circles and dead ends all over, and through the middle of everything runs the Thames River, spanned by a dozen bridges in central London, at least one of which — and sometimes more — will likely have to be crossed during a trip of any length through the city. And the erratic numbering system doesn't always tell you exactly where to find a particular address even when you've found the right street.

Thus the best advice for visitors is to forget about renting a car with a navigational system and instead rely on the city's cabbies. They're ubiquitous — some twenty-five thousand of them driving around in their big, black, boxy cars that are the automotive equivalent of sensible shoes — and they are astonishingly good at getting you from point A to point B in the most efficient way possible, taking into account not only the lengths of the various possible paths, but the time of day, the expected traffic, temporary roadwork and road closings, and any other details that might be relevant to the trip. Nor do points A and B have to be traditional street addresses. Suppose you'd like to revisit that funky little hat shop in Charing Cross whose name you don't quite recall — Lord's or Lear or something like that — but you do remember that there is a little shop next door that sells cupcakes. Well, that will be enough. Tell all that to your cabbie, and as soon as is automotively possible you will find yourself in front of Laird London, 23A New Row.

As you might imagine, given the challenges of finding one's way in London, not just anyone can be a cabbie. Indeed, to become a licensed London taxi driver one must pass a series of examinations that have been described as, collectively, the most difficult test in the world. The test is administered by Transport for London, and that agency describes "the Knowledge" — what a prospective driver must learn — as follows:

> To achieve the required standard to be licensed as an "All London" taxi driver you will need a thorough knowledge, primarily, of the area within a six-mile radius of Charing Cross. You will need to know: all the streets; housing estates; parks and open spaces; government offices and departments; financial and commercial centres; diplomatic premises; town halls; registry offices; hospitals; places of worship; sports stadiums and leisure centres; airline of-

> fices; stations; hotels; clubs; theatres; cinemas; museums; art galleries; schools; colleges and universities; police stations and headquarters buildings; civil, criminal and coroner's courts; prisons; and places of interest to tourists. In fact, anywhere a taxi passenger might ask to be taken.

That area within six miles of Charing Cross contains approximately twenty-five thousand streets. But a prospective cabbie must be familiar with more than just streets and buildings. Any landmark is fair game. According to a 2014 story about London taxi drivers in the *New York Times Magazine,* one prospective driver was asked about the location of a statue of two mice with a piece of cheese; the statue, on the façade of a building, was just one foot tall.

More to the point, prospective taxi drivers must demonstrate that they can get from one point in the city to another as efficiently as possible. Tests consist of a series of "runs" in which the examiner gives two points in London and the examinee must provide the precise location of each of the points and then describe the best route between them, turn by turn, naming each street in the sequence. Each run earns a numerical score based on its accuracy, and as the prospective driver accumulates points, the tests get harder and harder, with the endpoints becoming more obscure and the routes longer, more complicated, and more convoluted. Half or more of the prospective drivers end up dropping out, but those who stay with it and earn their licenses have internalized London to a degree that Google Maps, with its satellite images, camera cars, and unfathomable memory and processing power, can only vaguely approximate.

To master the Knowledge, prospective cabbies — who are known as "Knowledge boys" and, occasionally, "Knowledge girls" — spend years driving from place to place in London, making notes of what is where and how to get from here to there. The first step is to master a list of 320 runs in the guidebook provided to taxi-driver candidates. For a given run, a candidate will generally first figure out the shortest route by physically traveling the various possible routes, usually by motorbike, and then will explore the areas around the beginning and the end of the run. This means wandering around within a quarter mile or so of each of those places, taking notes on which buildings and which landmarks are in the vicinity. After having repeated this process 320 times, the prospective cabbie has accumulated a foundational set of 320 best routes around London and has also explored — and taken notes on — pretty much every bit of the core area within six miles of Charing Cross. It is a start, but successful candidates keep challenging themselves to determine the best routes for many other runs that are not on the list and to take note of buildings and landmarks that they might have missed before or that might have recently appeared. Indeed, even after passing all the tests and getting licensed, London taxi drivers continue to increase and hone their knowledge of London's streets.

The resulting memory and navigational skills are nothing short of astonishing, and so London taxi drivers have proved irresistible to psychologists interested in learning and, particularly, in the learning of navigational skills. By far the most in-depth studies of the cabbies — and the ones that have the most to tell us about how training affects the brain — have been carried out by Eleanor Maguire, a neuroscientist at University College London.

In one of her earliest works on the taxi drivers, published in 2000, Maguire used magnetic resonance imaging to look at the brains of sixteen male taxi drivers and compare them with the brains of fifty other males of similar ages who were not taxi drivers. She looked in particular at the hippocampus, that seahorse-shaped part of the brain involved in the development of memories. The hippocampus is particularly engaged by spatial navigation and in remembering the location of things in space. (Each person actually has two hippocampi, one on each side of the brain.) For instance, species of birds that store food in different places and thus must be able to remember the location of these various caches have relatively larger hippocampi than closely related birds that don't store food in different places. More to the point, the size of the hippocampus is quite flexible in at least some species of birds and can grow by as much as 30 percent in response to a bird's food-storing experiences. But would the same thing be true in humans?

Maguire found that a particular part of the hippocampus — the posterior, or rear, part — was larger in the taxi drivers than in the other subjects. Furthermore, the more time that a person had spent as a taxi driver, the larger the posterior hippocampi were. In another study that Maguire carried out a few years later, she compared the brains of London taxi drivers with London bus drivers. Like the taxi drivers, the bus drivers spent their days driving around London; the difference between them was that the bus drivers repeated the same routes over and over and thus never had to figure out the best way to get from point A to point B. Maguire found that the posterior hippocampi of the taxi drivers were significantly larger than the same parts of the brain in the bus drivers. The clear implication was that whatever was responsible for the difference in the size of the posterior hippocampi was not related to the driving itself but rather was related specifically to the navigational skills that the job required.

That still left one loose end, however: perhaps the taxi drivers in the studies had started out with larger posterior hippocampi that gave them an advantage in finding their way around London, and the extensive testing they went through was nothing more than a weeding-out process that zeroed in on those prospective drivers who were naturally better equipped to be able to learn their way around the maze that is London.

Maguire addressed this issue quite simply and powerfully: she followed a group of prospective taxi drivers from the time they started training for their licenses until the point at which all of them had either passed the tests and become licensed cabbies or else had given up and gone on to do something else. In particular, she recruited seventy-nine prospective drivers — all of them male — who were just starting training, as well as another thirty-one males of similar ages to serve as controls. When she scanned all their brains, she found no difference in the sizes of the posterior hippocampi between the prospective drivers and the controls.

Four years later she revisited the two groups of subjects. By this time forty-one of the trainees had become licensed London taxi drivers, while thirty-eight had stopped training or failed their tests. So at this point there were three groups to compare: the new taxi drivers who had learned enough about London's streets to pass the series of tests, the trainees who had not learned enough to pass, and the group who had not ever trained at all. Once again Maguire scanned their brains and calculated the size of the posterior hippocampi in each.

What she found would have been no surprise if she had been measuring biceps in bodybuilders, but she wasn't — she was measuring the sizes of different parts of the brain — and so the result was startling. The volume of the posterior hippocampi had gotten significantly larger in the group of trainees who had continued their training and had become licensed taxi drivers. By contrast, there was no change in the size of the posterior hippocampi among the prospective taxi drivers who had failed to become licensed (either because they simply stopped training or because they could not pass the tests) or among the subjects who had never had anything to do with the taxi training program. The years spent mastering the Knowledge had enlarged precisely that part of the brain that is responsible for navigating from one place to another.

Maguire's study, which was published in 2011, is perhaps the most dramatic evidence we have that the human brain grows and changes in response to intense training. Furthermore, the clear implication of her study is that the extra neurons and other tissue in the posterior hippocampi of the licensed cabbies underlie their increased navigational capabilities. You can think about the posterior hippocampi of a London taxi driver as the neural equivalent of the massively developed arms and shoulders of a male gymnast. Years of work on the rings and pommel horse and parallel bars and floor exercises have built muscles that are exquisitely suited for the sorts of movements he performs on those different pieces of apparatus — and, indeed, that make it possible for him to do all sorts of gymnastics moves that were simply not within his reach when he began training. The posterior hippocampi of the taxi drivers are equally "bulked up," but with brain tissue, not muscle fiber.

ADAPTABILITY

Until the first decade of the twenty-first century, most scientists would have flat out denied that something like what Maguire has seen in the brains of London cabbies was even possible. The general belief was that once a person reached adulthood, the wiring of his or her brain was pretty much fixed. Sure, everyone understood that there had to be tweaks here and there when you learned something new, but these were thought to be little more than the strengthening of some neural connections and the weakening of others, because the overall structure of the brain and its various neural networks were fixed. This idea went hand in hand with the belief that individual differences in abilities were due mainly to genetically determined differences in the brain's wiring and that learning was just a way of fulfilling one's genetic potential. One common metaphor depicted the brain as a computer: learning was like loading some data or installing new software — it allowed you to do some things you couldn't do before, but your ultimate performance would always be limited by such things as the number of bytes in your random-access memory (RAM) and the power of your central processing unit (CPU).

By contrast, the body's adaptability has always been easier to recognize, as we've noted. One of my favorite examples of physical adaptability involves pushups. If you're a relatively fit male in your twenties, you may be able to do 40 or 50; if you can do 100, you can impress your friends and probably win a few bets. So what might you guess is the world record for pushups — 500 or 1,000? In 1980 Minoru Yoshida of Japan did 10,507 pushups nonstop. After that, *Guinness World Records* stopped accepting submissions for the number of pushups done with no rest periods and switched to the most pushups performed in twenty-four hours with resting allowed. In 1993 Charles Servizio of the United States set what remains the world record in that category by doing 46,001 pushups in 21 hours and 21 minutes.

Or consider pull-ups. Even relatively fit guys can generally do only 10 or 15, although if you've really been working out, you may have worked your way up to 40 or 50. In 2014 Jan Kareš of the Czech Republic did 4,654 in twelve hours.

In short, the human body is incredibly adaptable. It is not just the skeletal muscles, but also the heart, the lungs, the circulatory system, the body's energy stores, and more — everything that goes into physical strength and stamina. There may be limits, but there is no indication that we have reached them yet.

From Maguire's work and that of others, we're now learning that the brain has a very similar degree and variety of adaptability.

Some of the earliest observations of this sort of adaptability — or "plasticity," as neuroscientists would say — appeared in studies of how the brains of blind or deaf people "rewire" themselves to find new uses for the parts of the brain that are normally dedicated to processing sights or sounds but that in these people have nothing to do. Most blind people cannot see because of problems with their eyes or optic nerve, but the visual cortex and other parts of the brain are still fully functional; they're just not getting any input from the eyes. If the brain actually were hardwired like a computer, these visual regions would sit forever idle. We now know, however, that the brain reroutes some of its neurons so that these otherwise-unused areas are put to work doing other things, particularly things related to the remaining senses, which blind people must rely on to get information about their surroundings.

To read, for example, the blind run their fingertips over the raised dots that make up the Braille alphabet. When researchers use MRI machines to watch the brains of blind subjects as they read words in Braille, one of the parts of the brain that they see lighting up is the visual cortex. In people with normal sight, the visual cortex would light up in response to input from the eyes, not the fingertips, but in the blind, the visual cortex helps them interpret the fingertip sensations they get from brushing over the groups of raised dots that make up the Braille letters.

Interestingly enough, it is not just otherwise-unused areas of the brain where rewiring occurs. If you practice something enough, your brain will repurpose neurons to help with the task even if they already have another job to do. Perhaps the most compelling evidence of this comes from an experiment done in the late 1990s, when a group of researchers examined the parts of the brain that controlled various fingers on the hands of a group of highly skilled Braille readers.

The subjects were three-fingered Braille readers — that is, they used their index fingers to read the patterns of dots that make up individual letters, their middle fingers to pick out the spaces between the letters, and their ring fingers to keep track of the particular line they were reading. The wiring in the part of the brain that controls the hands is normally set up so that each individual finger has a distinct part of the brain dedicated to it. This is what makes it possible for us to tell, for example, which fingertip is being touched by a pencil tip or a thumbtack without looking at our fingers. The subjects in the study were Braille instructors who used their fingers to read Braille several hours each day. What the researchers discovered was that this steady use of the three fingers had caused the areas of the brain devoted to each of those fingers to grow so much that those areas eventually overlapped. As a result, the subjects were exceptionally sensitive to touch on these fingers — they could detect a much gentler touch than sighted subjects — but they often couldn't tell which of the three fingers had been touched.

These studies of brain plasticity in blind subjects — and similar studies in deaf subjects — tell us that the brain's structure and function are not fixed. They change in response to use. It is possible to shape the brain — your brain, my brain, anybody's brain — in the ways that we desire through conscious, deliberate training.

Researchers are just beginning to explore the various ways that this plasticity can be put to work. One of the most striking results to date could have implications for anyone who suffers from age-related far-sightedness — which is just about everyone over the age of fifty. The study, which was carried out by American and Israeli neuroscientists and vision researchers, was reported in 2012. Those scientists assembled a group of middle-aged volunteers, all of whom had difficulty focusing on nearby objects. The official name of the condition is presbyopia, and it results from a problem with the eye itself, which loses elasticity in its lens, making it more difficult to focus well enough to make out small details. There is also an associated difficulty in detecting contrasts between light and dark areas, which exacerbates the difficulty in focusing. The consequences are a boon for optometrists and opticians and a bother for the over-fifty crowd, nearly all of whom need glasses to read or perform close-up work.

The researchers had their subjects come into the lab three or so times a week for three months and spend thirty minutes each visit training their vision. The subjects were asked to spot a small image against a background that was very similar in shade to the spot; that is, there was very little contrast between the image and the background. Spotting these images required intense concentration and effort. Over time the subjects learned to more quickly and accurately determine the presence of these images. At the end of three months the subjects were tested to see what size type they could read. On average they were able to read letters that were 60 percent smaller than they could at the beginning of the training, and every single subject had improved. Furthermore, after the training every subject was able to read a newspaper without glasses, something a majority of them couldn't do beforehand. They also were able to read faster than before.

Surprisingly, none of this improvement was caused by changes in the eyes, which had the same stiffness and difficulty focusing as before. Instead, the improvement was due to changes in the part of the brain that interprets visual signals from the eye. Although the researchers couldn't pinpoint exactly what those changes were, they believe that the brain learned to "de-blur" images. Blurry images result from a combination of two different weaknesses in vision — an inability to see small details and difficulties in detecting differences in contrast — and both of these issues can be helped by the image processing carried out in the brain, in much the same way that image-processing software in a computer or a camera can sharpen an image by such techniques as manipulating the contrast. The researchers who carried out the study believe that their training exercises taught the subjects' brains to do a better job of processing, which in turn allowed the subjects to discern smaller details without any improvement in the signal from the eyes.

CHALLENGING HOMEOSTASIS

Why should the human body and brain be so adaptable in the first place? It all stems, ironically enough, from the fact that the individual cells and tissues try to keep everything the same as much as possible.

The human body has a preference for stability. It maintains a steady internal temperature. It keeps a stable blood pressure and heart rate. It keeps the blood glucose levels and pH balance (acidity/alkalinity level) steady. It maintains a reasonably constant weight from day to day. None of these things are completely static, of course — pulse rate increases with exercise, for instance, and body weight goes up or down with overeating or dieting — but these changes are usually temporary, and the body eventually gets back to where it was. The technical term for this is "homeostasis," which simply refers to the tendency of a system — any sort of system, but most often a living creature or some part of a living creature — to act in a way that maintains its own stability.

Individual cells like stability as well. They maintain a certain level of water and also regulate the balance of positive and negative ions, particularly sodium and potassium ions, and various small molecules by controlling which ions and molecules stay and which exit through the cell membrane. More important to us is the fact that cells require a stable environment if they are to function effectively. If the surrounding tissues get too hot or too cold, if their fluid level moves too far outside of the preferred range, if the oxygen level drops too far, or if the energy supplies get too low, it damages the functioning of the cells. If the changes are too big for too long, the cells start to die.

Thus, the body is equipped with various feedback mechanisms that act to maintain the status quo. Consider what happens when you engage in some sort of vigorous physical activity. The contraction of muscle fibers causes the individual muscle cells to expend their supplies of energy and oxygen, which are replenished from nearby blood vessels. But now the level of oxygen and energy supplies in the bloodstream drops, which leads the body to take various measures in response. The breathing rate goes up to increase oxygen levels in the blood and to clear out more carbon dioxide. Various energy stores are converted into the sort of energy supply that the muscles can use and feed into the bloodstream. Meanwhile, blood circulation increases in order to better distribute the oxygen and energy supplies to those parts of the body that need them.

As long as the physical exercise is not so strenuous that it strains the body's homeostatic mechanisms, the exercise will do very little to prompt physical changes in the body. From the body's perspective, there is no reason to change; everything is working as it should.

It's a different matter when you engage in a sustained, vigorous physical activity that pushes the body beyond the point where the homeostatic mechanisms can compensate. Your body's systems and cells find themselves in abnormal states, with abnormally low levels of oxygen and various energy-related compounds, such as glucose, adenosine diphosphate (ADP), and adenosine triphosphate (ATP). The metabolism of the various cells can no longer proceed as usual, so there are different sets of biochemical reactions going on in the cells, producing an entirely different suite of biochemical products than the cell usually produces. The cells are not happy with this altered state of affairs, and they respond by calling up some different genes from the cells' DNA. (Most of the genes in the DNA of a cell are inactive at any given time, and the cell will "switch on" and "switch off" various genes, depending on what it needs at the time.) These newly activated genes will switch on or ramp up various biochemical systems within the cell, which will change its behavior in ways that are intended to respond to the fact that the cells and surrounding systems have been pushed out of their comfort zone.

The exact details of what goes on inside a cell in response to such stresses are extremely complicated, and researchers are only just now beginning to unravel them. For example, in one study on rats the scientists conducting the study counted 112 different genes that were turned on when the workload on a particular muscle in the rear legs of the rats was sharply increased. Judging by the particular genes that were switched on, the response included such things as a change in the metabolism of the muscle cells, changes in their structure, and a change in the rate at which new muscle cells were formed. The eventual result of all of these changes was a strengthening of the rats' muscles so that they could handle the increased workload. They had been pushed out of their comfort zone, and the muscles responded by getting strong enough to establish a new comfort zone. Homeostasis had been reestablished.

This is the general pattern for how physical activity creates changes in the body: when a body system — certain muscles, the cardiovascular system, or something else — is stressed to the point that homeostasis can no longer be maintained, the body responds with changes that are intended to reestablish homeostasis. Suppose, for example, that you begin a program of aerobic exercise — say, jogging three times a week for half an hour each time, keeping your heart rate at the recommended level of 70 percent of your maximum heart rate (which works out to something over 140 beats per minute for younger adults). The sustained activity will, among other things, lead to low levels of oxygen in the capillaries that supply your leg muscles. Your body will respond by growing new capillaries in order to provide more oxygen to the muscle cells in your legs and return them to their comfort zone.

This is how the body's desire for homeostasis can be harnessed to drive changes: push it hard enough and for long enough, and it will respond by changing in ways that make that push easier to do. You will have gotten a little stronger, built a little more endurance, developed a little more coordination. But there is a catch: once the compensatory changes have occurred — new muscle fibers have grown and become more efficient, new capillaries have grown, and so on — the body can handle the physical activity that had previously stressed it. It is comfortable again. The changes stop. So to keep the changes happening, you have to keep upping the ante: run farther, run faster, run uphill. If you don't keep pushing and pushing and pushing some more, the body will settle into homeostasis, albeit at a different level than before, and you will stop improving.

This explains the importance of staying just outside your comfort zone: you need to continually push to keep the body's compensatory changes coming, but if you push too far outside your comfort zone, you risk injuring yourself and actually setting yourself back.

This, at least, is the way the body responds to physical activity. Scientists know much less about how the brain changes in response to mental challenges. One major difference between the body and the brain is that the cells in the adult brain do not generally divide and form new brain cells. There are a few exceptions, such as in the hippocampus, where new neurons can grow, but in most parts of the brain the changes that occur in response to a mental challenge — such as the contrast training used to improve people's vision — won't include the development of new neurons. Instead, the brain rewires those networks in various ways — by strengthening or weakening the various connections between neurons and also by adding new connections or getting rid of old ones. There can also be an increase in the amount of myelin, the insulating sheath that forms around nerve cells and allows nerve signals to travel more quickly; myelination can increase the speed of nerve impulses by as much as ten times. Because these networks of neurons are responsible for thought, memories, controlling movement, interpreting sensory signals, and all the other functions of the brain, rewiring and speeding up these networks can make it possible to do various things — reading a newspaper without glasses, say, or quickly determining the best route from point A to point B — that one couldn't do before.

In the brain, the greater the challenge, the greater the changes — up to a point. Recent studies have shown that learning a new skill is much more effective at triggering structural changes in the brain than simply continuing to practice a skill that one has already learned. On the other hand, pushing too hard for too long can lead to burnout and ineffective learning. The brain, like the body, changes most quickly in that sweet spot where it is pushed outside — but not too far outside — its comfort zone.

SHAPING THE BRAIN

The fact that the human brain and body respond to challenges by developing new abilities underlies the effectiveness of purposeful and deliberate practice. The training of a London taxi driver or an Olympic gymnast or a violinist at a music academy is, in essence, a method of harnessing the adaptability of the brain and body to develop abilities that would otherwise be out of reach.

The best place to see this in action is in the development of musical ability. Over the past two decades brain researchers have studied in great detail how musical training affects the brain and how those effects in turn make possible extraordinary musical performance. The best known study was published in 1995 in the journal *Science*. Working with four German scientists, the psychologist Edward Taub at the University of Alabama at Birmingham recruited six violinists, two cellists, and a guitarist, all of whom were right-handed, to have their brains scanned. They also recruited six nonmusicians to serve as controls against whom the musicians would be compared. Taub wanted to see if there was any difference between the two groups in the areas of their brains that were devoted to controlling their fingers.

Taub was most interested in the fingers on the musicians' left hands. Playing the violin, cello, or guitar requires exceptional control of those fingers. The fingers move up and down the neck of the instrument and from string to string, sometimes at incredible speeds, and they must be placed with extreme accuracy. Furthermore, many of the sounds coaxed from the instruments, such as vibrato, involve some sliding or vibrating motion of a finger in place, which generally requires extensive practice to master. The left thumb has fewer responsibilities, mainly just providing pressure on the back of the neck, and the right hand generally has much less to do than the left — mostly just holding the bow for violinists and cellists and strumming or picking for guitarists. In short, most of a string player's training is aimed at improving control of the fingers on the left hand. The question Taub asked was, What effect will this have on the brain?

Taub's team used a magnetoencephalograph — a machine that maps out brain activity by detecting tiny magnetic fields in the brain — to determine which parts of the subjects' brains controlled which fingers. In particular, the experimenters would touch a subject's individual fingers and observe which parts of the brain responded to each touch. They found that the region of the brain controlling the left hand was significantly larger in the musicians than in the nonmusicians — and, in particular, that the brain regions controlling the fingers had taken over a section of the brain region that was normally devoted to the palm. Furthermore, the earlier a musician had started to play his or her instrument, the greater the expansion was. By contrast, the researchers found no difference between the musicians and nonmusicians in the size of the region controlling the fingers of the right hand.

The implication was clear: Years of practice on a stringed instrument had caused the area of the brain that controls the fingers of the left hand to gradually expand, resulting in a greater ability to control those fingers.

In the twenty years since that study, other researchers have expanded on the results and described a variety of ways that musical training affects brain structure and function. For example, the cerebellum — a part of the brain that plays an important role in controlling movements — is larger in musicians than in nonmusicians, and the more hours of training a musician has put in, the larger the cerebellum is. Musicians have more gray matter — the brain tissue that contains neurons — than nonmusicians do in various parts of the cortex, including the somatosensory region (touch and other senses), the superior parietal region (sensory input from the hands), and the premotor cortex (planning movements and guiding movements in space).

The details of exactly what happens to which region of the brain can be daunting to anyone who is not trained in neuroscience, but the big picture is clear: musical training modifies the structure and function of the brain in various ways that result in an increased capacity for playing music. In other words, the most effective forms of practice are doing more than helping you learn to play a musical instrument; they are actually increasing your *ability* to play. With such practice you are modifying the parts of the brain you use when playing music and, in a sense, increasing your own musical "talent."

Although less of this sort of research has been done in areas other than music, in every area that scientists have studied, the findings are the same: long-term training results in changes in those parts of the brain that are relevant to the particular skill being developed.

Some of these studies have focused on purely intellectual skills, such as mathematical ability. For example, the inferior parietal lobule has significantly more gray matter in mathematicians than in nonmathematicians. This part of the brain is involved in mathematical calculations and in visualizing objects in space, something that is important in many areas of math. It also happens to be a part of the brain that caught the attention of the neuroscientists who examined Albert Einstein's brain. They found that Einstein's inferior parietal lobule was significantly larger than average and that its shape was particularly unusual, which led them to speculate that his inferior parietal lobule may have played a crucial role in his ability to perform abstract mathematical thinking. Could it be that people like Einstein are simply born with beefier-than-usual inferior parietal lobules and thus have some innate capacity to be good at mathematical thinking? You might think so, but the researchers who carried out the study on the size of that part of the brain in mathematicians and nonmathematicians found that the longer someone had worked as a mathematician, the more gray matter he or she had in the right inferior parietal lobule — which would suggest that the increased size was a product of extended mathematical thinking, not something the person was born with.

A number of studies have examined skills that have both a mental and a physical component, such as playing music. One recent investigation looked at the brains of glider pilots versus nonpilots and found that the brains of the pilots had more gray area in several different regions, including the left ventral premotor cortex, the anterior cingulate cortex, and the supplementary eye field. These regions seem to be involved in such things as learning to control the stick that one uses to fly a glider, comparing the visual signals that one gets when flying with the body-balance signals that indicate the orientation of the glider, and controlling the movements of the eyes.

Even in the case of what we usually think of as purely "physical skills," such as swimming or gymnastics, the brain plays a major role because these activities require careful control of the body's movements, and research has found that practice produces brain changes. For instance, cortical thickness, a way of measuring the amount of gray matter in a brain area, is greater in competitive divers than in nondivers in three specific regions, all of which play a role in visualizing and controlling the movements of the body.

Although the specific details vary from skill to skill, the overall pattern is consistent: Regular training leads to changes in the parts of the brain that are challenged by the training. The brain adapts to these challenges by rewiring itself in ways that increase its ability to carry out the functions required by the challenges. This is the basic message that should be taken away from the research on the effects of training on the brain, but there are a few additional details that are worth noting.

First, the effects of training on the brain can vary with age in several ways. The most important way is that younger brains — those of children and adolescents — are more adaptable than adult brains are, so training can have larger effects in younger people. Because the young brain is developing in various ways, training at early ages can actually shape the course of later development, leading to significant changes. This is "the bent-twig effect." If you push a small twig slightly away from its normal pattern of growth, you can cause a major change in the ultimate location of the branch that grows from that twig; pushing on a branch that is already developed has much less effect.

One example of this effect is that adult pianists generally have more white matter in certain regions of the brain than nonmusicians do, with the difference being totally due to the amount of time spent practicing in childhood. The earlier a child gets started on the piano, the more white matter that pianist will have as an adult. So while you can learn to play the piano as an adult, it will not result in the same amount of extra white matter that would be produced if you learned to play as a child. At present no one knows what the practical implications of this are, but, generally speaking, more white matter leads to nerve signals being transmitted more quickly, so it seems likely that practicing the piano as a child will lead to certain neurological advantages that you just can't match with practice as an adult.

A second detail worth noting is that developing certain parts of the brain through prolonged training can come at a cost: in many cases people who have developed one skill or ability to an extraordinary degree seem to have regressed in another area. Maguire's study of the London taxi drivers provides perhaps the best example. At the end of the four years, when the trainees had either finished the course and become licensed drivers or had stopped trying, she tested her subjects' memory in two ways. One involved knowing the locations of various London landmarks, and at this the subjects who had become licensed drivers did far better than the rest of the subjects. The second was a standard test of spatial memory — remembering a complex figure after a thirty-minute delay — and on this the licensed drivers did much worse than the group who had never been trained to become taxi drivers. By contrast, the trainees who had dropped out scored about the same as the subjects who had never trained. Because all three groups scored equally well on this memory test at the start of the four-year period, the only explanation was that the licensed cabbies, by developing their memories of London streets, had done something to cause a decline in this other sort of memory. Although we don't know for sure what caused that, it seems likely that the intense training caused the trainees' brains to devote an increasingly large segment to this sort of memory, leaving less gray matter to devote to other sorts of memory.

Finally, the cognitive and physical changes caused by training require upkeep. Stop training, and they start to go away. Astronauts who spend months in space without gravity to work against come back to Earth and find it difficult to walk. Athletes who have to stop training because of a broken bone or torn ligament lose much of their strength and endurance in the limbs they cannot exercise. Similar things have been seen with athletes who have volunteered for studies in which they must lie in bed for a month or so. Strength fades. Speed diminishes. Endurance wilts.

And something similar is true with the brain. When Maguire studied a group of retired London taxi drivers, she found that they had less gray matter in their posterior hippocampi than did active taxi drivers, although they still had more than retired subjects who had never been taxi drivers. Once these taxi drivers had stopped using their navigational memory every day, the brain changes that had been the result of that work started to disappear.

BUILDING YOUR OWN POTENTIAL

Once we understand the adaptability of the brain and the body in this way, we start to think about human potential in an entirely different light, and it points us to an entirely different approach to learning.

Consider this: Most people live lives that are not particularly physically challenging. They sit at a desk, or if they move around, it's not a lot. They aren't running and jumping, they aren't lifting heavy objects or throwing things long distances, and they aren't performing maneuvers that require tremendous balance and coordination. Thus they settle into a low level of physical capabilities — enough for day-to-day activities and maybe even hiking or biking or playing golf or tennis on the weekends, but far from the level of physical capabilities that a highly trained athlete possesses. These "normal" people cannot run a mile in under five minutes or ten miles in under an hour; they cannot throw a baseball three hundred feet or hit a golf ball three hundred yards; they cannot do triple gainers off the high board or triple axels on ice skates or triple backflips in a gymnastics floor routine. These are the sorts of things that require far more practice than most people are willing to devote, but — and this is important — they are also the sorts of abilities that *can* be developed because the human body is so adaptable and responsive to training. The reason that most people don't possess these extraordinary physical capabilities isn't because they don't have the capacity for them, but rather because they're satisfied to live in the comfortable rut of homeostasis and never do the work that is required to get out of it. They live in the world of "good enough."

The same thing is true for all the mental activities we engage in, from writing a report to driving a car, from teaching a class to running an organization, from selling houses to performing brain surgery. We learn enough to get by in our day-to-day lives, but once we reach that point, we seldom push to go beyond good enough. We do very little that challenges our brains to develop new gray matter or white matter or to rewire entire sections in the way that an aspiring London taxi driver or violin student might. And, for the most part, that's okay. "Good enough" is generally good enough. But it's important to remember that *the option exists.* If you wish to become significantly better at something, you can.

And here is the key difference between the traditional approach to learning and the purposeful-practice or deliberate-practice approaches: The traditional approach is not designed to challenge homeostasis. It assumes, consciously or not, that learning is all about fulfilling your innate potential and that you can develop a particular skill or ability without getting too far out of your comfort zone. In this view, all that you are doing with practice — indeed, all that you can do — is to reach a fixed potential.

With deliberate practice, however, the goal is not just to reach your potential but to build it, to make things possible that were not possible before. This requires challenging homeostasis — getting out of your comfort zone — and forcing your brain or your body to adapt. But once you do this, learning is no longer just a way of fulfilling some genetic destiny; it becomes a way of taking control of your destiny and shaping your potential in ways that you choose.

The obvious next question is, What is the best way to challenge homeostasis and develop that potential? We will spend much of the rest of the book answering that question, but before we do that, we need to address an issue that we have glossed over in this chapter: What exactly are we trying to improve about our brains? It's pretty obvious what leads to improved physical abilities. If you build more and larger muscle fibers, you get stronger. If you improve your muscles' energy reserves, your lung capacity, your heart's pumping capacity, and the capacity of your circulatory system, you build your endurance. But what changes are you making in your brain as you train to be a musician, a mathematician, a taxi driver, or a surgeon? Surprisingly, there is a common theme to the changes in all of these areas, and understanding that is the key to understanding how people develop extraordinary abilities in any area of human performance with a mental component — which, when you think about it, is just about all of them. We discuss that next.

3

Mental Representations

ON APRIL 27, 1924, shortly before 2:00 p.m., the Russian grandmaster Alexander Alekhine settled into a comfortable leather chair at the front of a large room in the Hotel Alamac in New York City and prepared to play twenty-six of the best chess players in the area. The challengers sat at two long tables behind Alekhine. In front of each challenger was a chessboard on which that player's game with Alekhine would be played out. Alekhine could see none of the boards. Each time a player made a move, a runner would call out the board number and the move loudly enough that Alekhine could hear it, and then, once Alekhine announced his response, the runner would make Alekhine's move for him on the appropriate board.

Twenty-six boards, 832 individual pieces, and 1,664 individual squares to keep track of — all without taking notes or having any sort of memory aid — and yet Alekhine never stumbled. The demonstration went on for more than twelve hours, with a short break for dinner, and when the last game was finished, shortly after two o'clock in the morning, Alekhine had won sixteen, lost five, and played to a draw in five more.

This sort of chess game, in which one of the players — and sometimes both — cannot see the chessboard and must play from memory, is called "blindfold chess" even if there is no literal blindfold involved. Chess masters have been playing blindfolded for more than a thousand years, mostly as a way of showing off, although sometimes as a way of handicapping themselves when playing less-skilled opponents. Some of these old chess masters would even play blindfolded against two, three, or four opponents simultaneously, but it wasn't until the late nineteenth century that a few grandmasters began to get really serious with it, playing a dozen or more opponents at once. The current record is forty-six boards, set in 2011 by Marc Lang of Germany, with twenty-five wins, two losses, and nineteen draws. Nonetheless, Alekhine's exhibition in 1924 is still generally considered to be the most impressive simultaneous blindfold match ever because of the quality of his challengers and the number of his wins against such stiff competition.

Blindfold chess offers one of the most dramatic examples of what is possible to accomplish with purposeful practice. And learning a bit about blindfold chess can give us a clear idea of the sorts of neurological changes that arise from such practice.

THE ACCIDENTAL BLINDFOLD CHESS MASTER

Although Alekhine became interested in blindfold chess at an early age and played his first blindfold game when he was twelve, the vast majority of his training throughout his life was devoted not to blindfold chess but simply to chess.

Alekhine, who was born in October 1892, started playing chess when he was seven. By the time he was ten, he was playing in correspondence tournaments, and he would spend much of each day analyzing the positions in detail, even while he was in school. Because he could not bring a chessboard with him to class, he would write the position he was studying down on a piece of paper and puzzle over it during school hours. Once in an algebra class he suddenly stood up with a big smile on his face. "Well, have you solved it?" his teacher asked, referring to the algebra problem that the class had been assigned. "Yes," Alekhine responded, "I sacrifice the knight, the bishop moves, . . . and White wins!"

He first became interested in blindfold chess around the same time as he began playing in correspondence tournaments. The event that triggered Alekhine's interest in blindfold chess was a 1902 exhibition in Moscow by the U.S. chess champion Harry Nelson Pillsbury, who set a world record at that exhibition by playing twenty-two games simultaneously. As Alekhine would later tell it, his brother Alexei was one of Pillsbury's opponents that day, although the records of the match that we have today hold no indication that Alexei actually played. Either way, though, the demonstration made a serious impression on the young Alekhine, and a couple of years later he began to try blindfold chess himself. It was, he later wrote, a natural outgrowth of his habit of thinking about chess positions while he was in the classroom. At first he would sketch out positions and use the sketches to work out the best moves, but eventually he found that he could study the positions just as well without the diagrams — that he could hold the entire chessboard in his memory and move pieces around in his mind, trying various lines of play.

Over time Alekhine got to the point where he could play whole games in his mind with no need to look at a chessboard, and, as he got older, he began to play multiple blindfold games in the spirit of Pillsbury's demonstration. At sixteen he could do four or five blindfold games simultaneously, but he didn't pursue it any further, choosing instead to focus on improving his play in standard games. By this point it was clear to him that if he worked hard enough, he could become one of the best chess players in the world. And Alekhine, who had never lacked confidence in his abilities at chess, saw no reason to stop at the "one of the" part of that description. His goal was to be *the best* player, period — the World Chess Champion.

Alekhine was well on his way toward that goal when the First World War began, an interruption that rekindled his interest in blindfold chess. In early August 1914, Alekhine and many other chess masters were playing at a major tournament in Berlin when Germany declared war on both Russia and France. Many of the foreign chess players were interned, and Alekhine found himself in a prison with half a dozen of Russia's other best chess players — but with no chessboards. So until they were released back to Russia — for Alekhine it was more than a month — the chess masters entertained themselves by playing each other in games of blindfold chess.

Once back in Russia, Alekhine served with a Red Cross unit on the Austrian front, where in 1916 he suffered a serious injury to his spine and was captured by the Austrians. The Austrians kept him chained to a bed in a hospital for several months as his back healed. Once again he had little but chess to entertain himself, and he arranged for a number of local players to visit and play against him. During that time he regularly played blindfold games, perhaps to handicap himself against his less-skilled opponents. Once he returned to Russia, Alekhine again neglected blindfold chess until after he emigrated to Paris in 1921.

At this point Alekhine was actively seeking the World Chess Championship, and he needed some way to support himself during that quest. One of his few options was to give chess demonstrations, and so he began performing simultaneous blindfold matches. The first one he held in Paris was against twelve opponents, which were three or four more than he had ever played. At the end of 1923 he was in Montreal and decided to try to break the North American record for simultaneous blindfold chess games. The North American record at the time was twenty games, held by Pillsbury, so Alekhine played twenty-one. It went well, so he decided to go for the world record, which at the time was twenty-five games. That led to the demonstration at the Hotel Alamac. Over the coming years Alekhine would set the world record twice more — with twenty-eight games in 1925 and thirty-two games in 1933 — but he always contended that blindfold chess was little more than a way of bringing attention to the game of chess and, of course, to himself. It was never something that he made a special effort to develop, but rather something that grew out of his relentless effort to master the game and become the best in the world.

Alekhine eventually reached his goal, defeating José Raúl Capablanca in 1927 for the world championship. He held that title until 1935 and then again from 1937 to 1946, and many rankings place Alekhine among the ten best chess players of all time. But when people rank the greatest blindfold chess players of all time, Alekhine's name is generally at the very top of the list — even though blindfold chess was never his main focus.

If we look at the overall history of blindfold chess, we find that this same thing is true for most blindfold chess players throughout history. They worked to become chess masters, and they found themselves, with little or no additional effort, able to play blindfolded.

At first glance, the way that so many grandmasters develop the ability to play blindfold chess might seem nothing more than an artifact, an interesting footnote to the history of chess. But if you look more closely, you find that this connection is actually a clue pointing toward the particular mental processes that set chess masters apart from chess novices and make possible their incredible ability to analyze chess positions and zero in on the best moves. Furthermore, the same sorts of highly developed mental processes are seen in expert performers in every field and hold the key to understanding their extraordinary abilities.

Before we delve into this, however, let us take a quick detour to examine in more detail the sort of memory that chess experts have for pieces arranged on a chessboard.

THE SECRET TO WINNING CHESS

Beginning in the early 1970s researchers sought to understand how grandmasters remember chess positions with such accuracy. The earliest studies were done by my mentor, Herb Simon, working with Bill Chase, who would later be my collaborator in my studies of Steve Faloon's digit-span memory.

It was already known that grandmasters, given just a few seconds to study a chessboard as it appears in the middle of a game, will remember correctly the position of most of the pieces and be able to reproduce the most important areas of the board almost perfectly. This ability seemed to defy the well-known limits of short-term memory. By contrast, someone who is just beginning to play chess can remember the positions of only a handful of pieces and can't come close to reconstructing the arrangement of pieces on the board.

Herb and Bill asked a simple question: Are chess experts recalling the position of each piece, or are they actually remembering patterns, where the individual pieces are seen as part of a larger whole? To answer that question Herb and Bill carried out a simple but effective experiment. They tested a national-level chess player (i.e., a chess master), a mid-range chess player, and a chess novice on two types of boards, one that had the pieces arranged in a pattern taken from a real chess game, and the other with a random jumble of pieces of the sort that made no chess sense at all.

When shown chessboards with a dozen to two dozen pieces arranged in a pattern from the middle or the end of a chess game, the master could remember the positions of about two-thirds of the pieces after five seconds of study, the novice could remember only about four, and the mid-range player was somewhere in the middle. When shown chessboards with the pieces arrayed randomly, the novice player did somewhat worse — only about two pieces correct. No surprise there. What was surprising, however, was that neither the mid-range player nor the chess master did much better than the novice in remembering the positions of pieces arranged randomly on a board. They too got only about two or three pieces right. The experienced players' advantage had disappeared. More recent studies of large groups of chess players have reiterated the original findings.

Something very similar has been shown with verbal memory. If you ask someone to recall a seemingly random assortment of words verbatim, starting with the first word — "was smelled front that his the peanuts he good hunger eating barely woman of so in could that him contain" — the average person will remember only the first six of those words. If, however, you read the same words rearranged into a sentence that makes clear sense — "The woman in front of him was eating peanuts that smelled so good that he could barely contain his hunger" — some adults will remember all of the words in perfect order, and most people will remember most of the sentence. What's the difference? The second arrangement carries meaning that allows us to make sense of the words using preexisting "mental representations." They're not random; they mean something, and meaning aids memory. Similarly, chess masters don't develop some incredible memory for where individual pieces sit on a board. Instead, their memory is very context-dependent: it is only for patterns of the sort that would appear in a normal game.

The ability to recognize and remember meaningful patterns arises from the way chess players develop their abilities. Anyone who is serious about developing skills on the chessboard will do it mainly by spending countless hours studying games played by the masters. You analyze a position in depth, predicting the next move, and if you get it wrong, you go back and figure out what you missed. Research has shown that the amount of time spent in this sort of analysis — not the amount of time spent playing chess with others — is the single most important predictor of a chess player's ability. It generally takes about ten years of this sort of practice to reach the level of grandmaster.

These years of practice make it possible for chess players to recognize patterns of chess pieces — not just their positions, but the interactions among them — at a glance. They are old friends. Bill Chase and Herb Simon called these patterns "chunks," and the important thing about them is that they are held in long-term memory.

Simon estimated that by the time a chess player becomes a master, he or she has accumulated some fifty thousand of these chunks. A master who examines a chess position sees a collection of chunks that are interacting with other chunks in still other patterns. Research has shown that these chunks are organized hierarchically, with groups of chunks arranged into higher-level patterns. The hierarchy is analogous to the organizational structure of a business or other large institution, with individuals organized into teams, which are organized into units, which are organized into departments, and so on, with the higher-level pieces being more abstracted and further from the bottom level where the real action takes place (which, in the case of the chess example, is the level of the individual chess pieces).

The way that grandmasters process and make sense of chess positions is an example of a mental representation. It is their way of "seeing" the board, and it's quite different from how a novice would see the same board.

When asked what they see when they are mentally examining a chess position, grandmasters do not talk about visualizing the physical chess pieces on a board as they would if they were relying on some sort of "photographic memory" of the position. This would be a "bottom-level" representation. Instead, their descriptions are much more vague, sprinkled with such terms as "lines of force" and "power." A key thing about these representations is that they allow a chess player to encode the positions of pieces on the board in a much more efficient way than simply remembering which piece is on which square. This efficient encoding underlies a master's ability to glance at a chessboard and remember the positions of most of the pieces and, in particular, the ability to play blindfold chess.

Two other features of these representations are worth noting, for they are examples of themes that will appear again and again as we explore the broader world of mental representations.

First, the mental representations are more than just ways of encoding positions. They allow a chess master to glance at a game in progress and get an immediate sense of which side has the advantage, which directions the game might take, and what a good move or moves might be. This is because the representations include, in addition to the positions of the pieces and the interactions between them, the various weaknesses and strengths of the two players' positions and moves that are likely to be effective in such positions. One of the things that most clearly sets grandmasters apart from novices or mid-range players is their ability to devise much better potential moves when they first examine a position.

The second notable characteristic of these mental representations is that while a chess master will initially analyze a position in terms of general patterns — which is enough when playing a lesser opponent — the representations also allow the master to zero in on individual pieces and mentally move them around the board to see how such moves would change the patterns. So the master can quickly examine strings of possible moves and countermoves in great detail, looking for the particular move that will offer the best chance of winning. In short, while the mental representations give masters a view of the forest that novices lack, they also allow masters to zero in on the trees when necessary.

MENTAL REPRESENTATIONS

Mental representations aren't just for chess masters; we all use them constantly. A mental representation is a mental structure that corresponds to an object, an idea, a collection of information, or anything else, concrete or abstract, that the brain is thinking about. A simple example is a visual image. Mention the *Mona Lisa,* for instance, and many people will immediately "see" an image of the painting in their minds; that image is their mental representation of the *Mona Lisa.* Some people's representations are more detailed and accurate than others, and they can report, for example, details about the background, about where Mona Lisa is sitting, and about her hairstyle and her eyebrows.

A somewhat more complex example of a mental representation is a word — *dog,* for example. Suppose you've never heard of a dog and never seen anything like it. Perhaps you've grown up in some isolated place — a desert island, say — where there are no four-legged animals of any sort, only birds and fish and insects. When you're first introduced to the concept of *dog,* it is all just isolated data, and the word *dog* doesn't really mean much to you; it is just a label for this collection of disconnected knowledge. Dogs are furry, they have four legs, they are meat eaters, they run in packs, the little ones are called puppies, they can be trained, and so on. Gradually, however, as you spend time around dogs and start to understand them, all this information becomes integrated into one holistic concept that is represented by the word *dog.* Now when you hear that word, you don't have to search your memory banks to remember all the various details about dogs; instead, all that information is immediately accessible. You have added *dog* not only to your vocabulary but to your set of mental representations.

Much of deliberate practice involves developing ever more efficient mental representations that you can use in whatever activity you are practicing. When Steve Faloon was training to improve his ability to remember long strings of digits, he developed increasingly sophisticated ways to encode those digits mentally — that is, he created mental representations. When London taxi trainees are learning to navigate efficiently from every point A to every point B in the city, they do it by developing increasingly sophisticated mental maps of the city — that is, by making mental representations.

Even when the skill being practiced is primarily physical, a major factor is the development of the proper mental representations. Consider a competitive diver working on a new dive. Much of the practice is devoted to forming a clear mental picture of what the dive should look like at every moment and, more importantly, what it should feel like in terms of body positioning and momentum. Of course, the deliberate practice will also lead to physical changes in the body itself — in divers, the development of the legs, abdominal muscles, back, and shoulders, among other body parts — but without the mental representations necessary to produce and control the body's movements correctly, the physical changes would be of no use.

A key fact about such mental representations is that they are very "domain specific," that is, they apply only to the skill for which they were developed. We saw this with Steve Faloon: the mental representations he had devised to remember strings of digits did nothing to improve his memory for strings of letters. Similarly, a chess player's mental representations will give him or her no advantage over others in tests involving general visuospatial abilities, and a diver's mental representations will be useless for basketball.

This explains a crucial fact about expert performance in general: there is no such thing as developing a general skill. You don't train your memory; you train your memory for strings of digits or for collections of words or for people's faces. You don't train to become an athlete; you train to become a gymnast or a sprinter or a marathoner or a swimmer or a basketball player. You don't train to become a doctor; you train to become a diagnostician or a pathologist or a neurosurgeon. Of course, some people do become overall memory experts or athletes in a number of sports or doctors with a general set of skills, but they do so by training in a number of different areas.

Because the details of mental representations can differ dramatically from field to field, it's hard to offer an overarching definition that is not too vague, but in essence these representations are preexisting patterns of information — facts, images, rules, relationships, and so on — that are held in long-term memory and that can be used to respond quickly and effectively in certain types of situations. The thing all mental representations have in common is that they make it possible to process large amounts of information quickly, despite the limitations of short-term memory. Indeed, one could define a mental representation as a conceptual structure designed to sidestep the usual restrictions that short-term memory places on mental processing.

The best example of this that we have seen is Steve Faloon's ability to recall as many as eighty-two digits when only seven or eight digits would have been possible if he'd had to rely on short-term memory alone. He did it by encoding the digits he was hearing, three or four at a time, into meaningful memories in his long-term memory and then associating these memories with the retrieval structure, which allowed him to remember which digit group followed which. To do all this he needed mental representations not just for the three- and four-digit groups of numbers that he was holding on to but also for the retrieval structure itself, which he visualized as a sort of two-dimensional tree with the three- and four-digit groups placed at the ends of the individual branches.

But memorizing lists of things is just the simplest example of how short-term memory comes into play in our lives. We constantly have to hold on to and process many pieces of information simultaneously: the words in a sentence whose meaning we are figuring out, the positions of the pieces on a chessboard, or the different factors we must take into account when driving a car, such as our own speed and momentum, the positions and speeds of other vehicles, the road conditions and visibility, where our foot must be to hit the accelerator or brake, how much force to apply to the pedals, how quickly to turn the steering wheel, and so on. Any relatively complicated activity requires holding more information in our heads than short-term memory allows, so we are always building mental representations of one sort or another without even being aware of it. Indeed, without mental representations we couldn't walk (too many muscle movements to coordinate), we couldn't talk (ditto on the muscle movements, plus no understanding of the words), we couldn't live any sort of human life.

So everyone has and uses mental representations. What sets expert performers apart from everyone else is the quality and quantity of their mental representations. Through years of practice, they develop highly complex and sophisticated representations of the various situations they are likely to encounter in their fields — such as the vast number of arrangements of chess pieces that can appear during games. These representations allow them to make faster, more accurate decisions and respond more quickly and effectively in a given situation. This, more than anything else, explains the difference in performance between novices and experts.

Consider how professional baseball players are able to consistently hit balls that are pitched toward them at speeds that may exceed ninety miles per hour — something that's impossible for anyone who hasn't spent years training in that particular skill. These hitters have just a split second to decide whether to swing and, if so, where to swing. They have no better eyesight than an average person, and their reflexes are no faster. What they have is a set of mental representations developed through years of hitting pitches and getting immediate feedback about their expectations concerning a pitch. These representations enable them to quickly recognize what sort of pitch is coming and where it will likely be when it reaches them. As soon as they see the pitcher's arm come around and the ball leave his hand, they have a very good idea — without having to do any sort of conscious calculations — whether it will be a fastball, slider, or curve and approximately where it's heading. In essence, they've learned to read the pitcher's delivery, so they have less need to actually see how the ball travels before determining whether and where to swing the bat. The rest of us, who are illiterate where pitching is concerned, simply can't make these decisions before the ball arrives in the catcher's mitt.

So here is a major part of the answer to the question we asked at the end of the last chapter: What exactly is being changed in the brain with deliberate practice? The main thing that sets experts apart from the rest of us is that their years of practice have changed the neural circuitry in their brains to produce highly specialized mental representations, which in turn make possible the incredible memory, pattern recognition, problem solving, and other sorts of advanced abilities needed to excel in their particular specialties.

The best way to understand exactly what these mental representations are and how they work is, fittingly enough, to develop a good mental representation of the concept *mental representation.* And just as was the case with *dog,* the best way to develop a mental representation of mental representations is to spend a little time getting to know them, stroking their fur, patting their little heads, and watching as they perform their tricks.

RECOGNIZING AND RESPONDING TO PATTERNS

In pretty much every area, a hallmark of expert performance is the ability to see patterns in a collection of things that would seem random or confusing to people with less well developed mental representations. In other words, experts see the forest when everyone else sees only trees.

This is perhaps most obvious in team sports. Take soccer, for instance. You have eleven players on a side moving around in a way that to the uninitiated seems a swirling chaos with no discernible pattern beyond the obvious fact that some players are drawn to the soccer ball whenever it comes near. To those who know and love the game, however, and particularly to those who play the game well, this chaos is no chaos at all. It is all a beautifully nuanced and constantly shifting pattern created as the players move in response to the ball and the movements of the other players. The best players recognize and respond to the patterns almost instantaneously, taking advantage of weaknesses or openings as soon as they appear.

To study this phenomenon, I and two colleagues, Paul Ward and Mark Williams, investigated how well soccer players can predict what's coming next from what has already happened on the field. To do this we showed them videos of real soccer matches and suddenly stopped the video when a player had just received the ball. Then we asked our subjects to predict what would happen next. Would the player with the ball keep it, attempt a shot at the goal, or pass the ball to a teammate? We found that the more accomplished players were much better at deciding what the player with the ball should do. We also tested the players' memory for where the relevant players were located and in what directions they were moving by asking them to recall as much as they could from the last frame of the video before it was hidden from them. Again, the better players outperformed the weaker ones.

We concluded that the advantage better players had in predicting future events was related to their ability to envision more possible outcomes and quickly sift through them and come up with the most promising action. In short, the better players had a more highly developed ability to interpret the pattern of action on the field. This ability allowed them to perceive which players' movements and interactions mattered most, which allowed them to make better decisions about where to go on the field, when to pass the ball and to whom, and so on.

Something very similar is true for football, although it is mainly the quarterback who needs to develop mental representations of events on the field. This explains why the most successful quarterbacks are generally the ones who spend the most time in the film room, watching and analyzing the plays of their own team and their opponents. The best quarterbacks keep track of what's happening everywhere on the field, and after the game they can generally recall most of the game's plays, providing detailed descriptions of the movements of many players on each team. More importantly, effective mental representations allow a quarterback to make good decisions quickly: whether to pass the ball, whom to pass to, when to pass, and so on. Being able to make the right decision a tenth of a second faster can be the difference between a good play and a disastrous one — between, say, a completed pass and an interception.

Another important fact about mental representations was shown in a 2014 study by German researchers who looked at indoor rock climbing. In this sport, which is designed to mimic and serve as training for outdoor rock climbing, one must climb a vertical wall using various handholds. These holds require different kinds of grips, including an open grip, a pocket grip, a sideways pull, and a crimp grip. For each grip, the climber must position his or her hands and fingers differently. If you use the wrong grip on a handhold, you're more likely to fall.

Using standard psychological techniques, the researchers examined what was going on in the brains of climbers when they surveyed the various holds. They first noted that, unlike novices, experienced climbers automatically identified each hold according to the type of grip it required. In their mental representations of the various holds, all of the holds that required a crimp grip, for instance, were put together in one group and were distinguished from the holds that required, say, a pocket grip, which were themselves mentally put into a separate group. This grouping was done unconsciously, just as you can look at a poodle and Great Dane and know instantly they are both members of the same category without ever actually saying to yourself, "Those are both dogs."

In other words, experienced climbers had developed mental representations of holds that allowed them to know without conscious thought what sort of grip was required for each hold they saw. Furthermore, the researchers found that when experienced climbers saw a particular hold, their brains sent a signal to their hands preparing them to form the corresponding grip — again, without conscious thought. The inexperienced climbers had to consciously figure out the appropriate grip for each hold. The ability of experienced climbers to automatically analyze holds using a mental representation allows them to climb more quickly and with less chance of falling. Again, better mental representations lead to better performance.

MAKING SENSE OF INFORMATION

For the experts we just described, the key benefit of mental representations lies in how they help us deal with information: understanding and interpreting it, holding it in memory, organizing it, analyzing it, and making decisions with it. The same is true for all experts — and most of us are experts at something, whether we realize it or not.

For instance, most everyone reading this right now is an "expert" in reading, and to get to that level you had to develop certain mental representations. It began with learning the correspondence between letters and sounds. At that point reading was a matter of laboriously sounding out each word, letter by letter. With practice, you began to recognize entire words by themselves. C-A-T became simply *cat,* thanks to a mental representation that encoded the pattern of the letters in that word and associated that pattern with both the sound of the word and the idea of a small, furry animal that meows and often doesn't get along well with dogs. Along with the mental representations for the words, you developed a variety of other representations that are essential in reading. You learned how to recognize the beginning and ending of a sentence so that you could break up the strings of words into chunks that had individual meaning, and you learned that certain things that looked like they signaled the end of a sentence — Mr., Ms., Dr., and so on — usually did not. You internalized various patterns that allow you to infer the meanings of words you have never seen before and to use context to make sense of things where a word is misspelled or misused or left out altogether. And now when you read, you do all of this unconsciously, the mental representations churning away under the surface, unnoticed but essential.

While almost all of you reading this are experts in reading, in the sense that you are fully capable of recognizing the marks on the page as corresponding to words and sentences in your language, some of you will be more expert than others in the task of understanding and assimilating the information contained in this book. And, again, this is related to how well your mental representations allow you to overcome the limitations of short-term memory and retain what you are reading.

To see why, consider what happens when you test a group of subjects by having them read a newspaper article on something a bit specialized — say, a football or baseball game — and then quiz them to see how much of it they remember. You might guess that the results would depend mainly on the subjects' general verbal ability (which is closely related to IQ), but you'd be wrong. Studies have shown that the key factor determining a person's comprehension of a story about a football or baseball game is how much that person already understands about the sport.

The reason is straightforward: If you don't know much about the sport, then all of the details you read are essentially a bunch of unrelated facts, and remembering them is not much easier than remembering a list of random words. But if you understand the sport, you've already established a mental structure for making sense of it, organized the information, and combined it with all the other relevant information you've already assimilated. The new information becomes part of an ongoing story, and as such it moves quickly and easily into your long-term memory, allowing you to remember far more of the information in an article than you could if you were unfamiliar with the game it describes.

The more you study a subject, the more detailed your mental representations of it become, and the better you get at assimilating new information. Thus a chess expert can look at a series of moves in chess notation that are gibberish to most people — 1. e4 e5 2. Nf3 Nc6 3. Bb5 a6 . . . — and follow and understand an entire game. Similarly, an expert musician can look at a musical score for a new composition and know what it will sound like before ever playing it. And if you are a reader who is already familiar with the concept of *deliberate practice* or with the broader area of the psychology of learning, you will likely find it easier than other readers to assimilate the information in this book.

Either way, reading this book and thinking about the topics I'm discussing will help you create new mental representations, which will in turn make it easier for you to read and learn more about this subject in the future.

FINDING AN ANSWER

Every so often the *New York Times* publishes a column called "Think Like a Doctor" by Lisa Sanders, a doctor and author. Each column poses a medical mystery, a real case that initially puzzled the clinicians who encountered it — the newspaper version of an episode of *House M.D.* Sanders gives the readers enough information to solve it themselves — assuming they have all the other tools they need, such as medical knowledge and the ability to reason from symptoms to diagnosis — and then invites answers. In a later column she reveals the correct answer, explains how the original doctors reached that answer, and announces how many readers got it right. These columns always draw hundreds of reader responses — and only a few correct ones.

For me the most fascinating thing about the column is not the medical mysteries or their solutions but rather the insights that the column offers into the diagnostic thought process. A doctor making a diagnosis, particularly in a complex case, is given a large number of facts about the patient's condition and must absorb those facts and then combine them with relevant medical knowledge to come to a conclusion. This doctor must do at least three different things: assimilate facts about the patient, recall relevant medical knowledge, and use the facts and medical knowledge to identify possible diagnoses and choose the right one. For all of these activities, a more sophisticated mental representation makes the process faster and more efficient — and sometimes makes it possible, period.

To see how this works, I'll borrow one of Sanders's medical mysteries — one that only a handful of readers solved correctly, out of more than two hundred who sent in answers. A thirty-nine-year-old male police officer came to his doctor complaining of an intense earache — it felt like a knife in his ear — and noting that his right pupil was smaller than his left. He had had the earache once before and had visited an urgent-care center, where he was diagnosed with an infection and given a prescription for antibiotics. When it got better in a couple of days, he thought nothing more about it, but the earache reappeared two months later, and this time the antibiotics did no good. The doctor thought it was probably just a sinus infection, but because of the issue with the pupil, the patient was referred to an eye doctor. That doctor couldn't make a diagnosis and referred the patient to a specialist. The specialist, a neuro-opthalmologist, immediately recognized the small pupil as a symptom of a particular syndrome but had no idea what might cause that syndrome in an otherwise healthy man — and how it all might be related to the intense ear pain. So he asked a number of questions: Have you felt weakness anywhere? What about numbness or tingling? Have you been lifting weights recently? When the patient replied that he had been lifting weights for several months, the doctor asked one more question: Had he experienced any serious pain in his head or neck after lifting? Yes, he'd had an intense post-workout headache a couple of weeks earlier. The doctor could finally figure out what was wrong.

At first, the essential step in solving this mystery might seem to be recognizing what syndrome could cause one pupil to be smaller than the other, but this was actually pretty straightforward: it required having learned about that syndrome at some point and being able to recall its symptoms. It is called "Horner's syndrome," and it is caused by damage to a nerve that runs behind the eye. The damage hurts the eye's ability to dilate and often limits the movement of the eyelid covering that eye — and, indeed, when the specialist looked closely, he could see that that eyelid was not opening fully. Several readers identified Horner's syndrome correctly but were stymied by how that might be connected to the ear pain.

In this particular sort of challenge — piecing together a number of clues — the mental representations of expert doctors come to the fore. A doctor diagnosing a patient with a complex set of symptoms must take in a great deal of information without knowing ahead of time which is most relevant and which might be red herrings. It's impossible to assimilate all that information as random facts — the limitations of short-term memory will not allow it — so it must be understood against the background of the relevant medical knowledge. But what is relevant? Before a diagnosis is made, it is difficult to know what the various bits of clinical information might imply and what sorts of medical conditions they might be related to.

Medical students, whose mental representations of diagnostic medicine are still rudimentary, tend to associate symptoms with the particular medical conditions that they're familiar with and jump quickly to conclusions. They fail to generate multiple options. Even many less experienced doctors do the same thing. Thus, when the police officer went to the urgent-care clinic complaining of an earache, the doctor there assumed the problem was an infection of some sort — which would have been the correct answer in most cases — and didn't worry about the seemingly irrelevant fact that one of the patient's pupils was acting up.

Unlike medical students, expert diagnosticians have built sophisticated mental representations that let them consider a number of different facts at once, even facts that at first might not seem germane. This is a major advantage of highly developed mental representations: you can assimilate and consider a great deal more information at once. Research on expert diagnosticians has found that they tend to see symptoms and other relevant data not as isolated bits of information but as pieces of larger patterns — in much the same way that grandmasters see patterns among chess pieces rather than a random assortment of pieces.

And just as chess masters' representations allow them to quickly generate a number of possible moves and then zero in on the best one, experienced diagnosticians come up with a number of possible diagnoses and then analyze the various alternatives to select the most likely one. Of course, the doctor may ultimately decide that none of the options work, but the process of reasoning through each of them may well have led to still other possibilities. This ability to generate a number of likely diagnoses and carefully reason through them distinguishes expert diagnosticians from the rest.

The solution to the medical mystery described in the *New York Times* required precisely that sort of approach: first come up with possible explanations for why a patient should have both Horner's syndrome and a knifelike pain in the ear, and then analyze each possibility to find the right answer. Stroke was one possibility, but the patient had nothing in his background that indicated he might have had a stroke. Shingles could also produce the patient's two symptoms, but he had none of the usual signs of shingles such as blisters or a rash. A third possibility was a tear in the wall of the carotid artery, which runs right alongside the nerve affected in Horner's and also passes near the ear. A slight tear in the artery can allow blood to leak through the inner walls of the artery, causing a bulge in its outer wall, which can press on the nerve to the face and, in rare cases, also press on a nerve to the ear. With this in mind, the specialist asked the patient questions about lifting weights and headaches. It is known that weightlifting can sometimes tear the carotid artery, and such a tear would normally be associated with some sort of headache or neck pain. When the patient answered yes, the specialist decided that a tear in the carotid artery was the most likely diagnosis. An MRI scan verified that diagnosis, and the patient was put on blood thinners to prevent the formation of a blood clot and was told to avoid any sort of exertion for the several months that it would take the blood vessel to heal.

The key to the successful diagnosis wasn't merely having the necessary medical knowledge, but having that knowledge organized and accessible in a way that allowed the doctor to come up with possible diagnoses and to zero in on the most likely. The superior organization of information is a theme that appears over and over again in the study of expert performers.

This is true even for something as mundane as insurance sales. A recent study examined knowledge about multiline insurance (life, home, auto, and commercial) in 150 agents. Not surprisingly, the highly successful agents — as determined by their sales volumes — knew more about the various insurance products than the less successful agents. But more to the point, researchers found that the highly successful agents had much more complex and integrated "knowledge structures" — what we're calling mental representations — than the less successful agents. In particular, the better agents had much more highly developed "if . . . then" structures: if these things are true about a client, then say this or do that. Because their insurance knowledge was better organized, the best agents could figure out what to do more quickly and more accurately in any given situation, and this made them much more effective agents.

PLANNING

Before experienced rock climbers begin a climb, they will look over the entire wall and visualize the path they are going to take, seeing themselves moving from hold to hold. This ability to create a detailed mental representation of a climb before embarking on it is something that only comes with experience.

More generally, mental representations can be used to plan a wide variety of areas, and the better the representation, the more effective the planning.

Surgeons, for example, will often visualize an entire surgery before making the first incision. They use MRIs, CT scans, and other images to take a look inside the patient and identify potential trouble spots, then they devise a plan of attack. Developing such mental representations of a surgery is one of the most challenging — and most important — things that a surgeon can do, and more experienced surgeons generally create more sophisticated and more effective representations of these procedures. The representations not only guide the surgery, but they also serve to provide a warning when something unexpected and potentially dangerous happens in the surgery. When an actual surgery diverges from the surgeon's mental representation, he or she knows to slow down, rethink the options, and, if necessary, formulate a new plan in response to the new information.

Relatively few of us climb rocks or perform surgeries, but almost everyone writes, and the process of writing offers us an excellent example of how mental representations can be used in planning. I myself have become quite familiar with this arena over the past couple of years as I've worked on *Peak,* and many of the people reading this book will have done some writing lately as well, whether it has been a personal letter or a business memo, a blog post or a book.

There has been quite a bit of research into the representations people use when writing, and the research has demonstrated a profound difference between the methods used by expert writers and those used by novices. Consider, for example, the answer that a sixth-grader offered when asked about the strategy he used in writing an essay:

> I have a whole bunch of ideas and write down till my supply of ideas is exhausted. Then I might try to think of more ideas up to the point where you can't get any more ideas that are worth putting down on paper and then I would end it.

This approach is actually pretty typical, not just for sixth-graders but for many people who don't write for a living. The representation of the writing is simple and direct: there's a topic and there are various thoughts that the writer has on the topic, often loosely organized by relevance or importance, but sometimes by category or some other pattern. A slightly more sophisticated representation might include some sort of introduction at the beginning and a conclusion or summary at the end, but that's about it.

This approach to writing has been called "knowledge telling" because it is little more than telling the reader whatever comes into your head.

Expert writers do it very differently. Consider how my coauthor and I put this book together. First we had to figure out what we wanted the book to do. What did we want readers to learn about expertise? What concepts and ideas were important to introduce? How should a reader's ideas about training and potential be changed by reading this book? Answering questions like these gave us our first rough mental representation of the book — our goals for it, what we wanted it to accomplish. Of course, as we worked more and more on the book, that initial image evolved, but it was a start.

Next we started sketching out how we would accomplish our goals for the book. What general topics did we need to cover? Obviously we needed to explain what deliberate practice is. How would we do that? Well, first we would need to explain how people normally practice and the limitations of that approach, and then we would discuss purposeful practice, and so on. At that point we were envisioning various approaches we could use to reach our goals for the book and weighing them, seeing which options seemed best.

As we made our choices, we gradually honed our mental representation of the book until we had something that seemed to meet all of our goals. The simplest way to imagine our mental representation at this stage is to think back to the old outlining technique you learned back in junior high English class. We prepared an outline of chapters, each focused on a particular topic and covering various aspects of that topic. But the representation of the book that we had created was far richer and more complex than a simple outline. We knew, for instance, why each piece was there and what we wanted to accomplish with it. And we had a clear idea of the book's structure and logic — why one topic followed another — and the interconnections among the various pieces.

We found that this process also forced us to think carefully about how we conceptualized deliberate practice ourselves. We started off with what seemed to be a clear idea of deliberate practice and how to explain it, but as we tried to describe it briefly in a nontechnical way, sometimes we found that it just wasn't working as well as we would have liked. That would lead us to rethink the best way to explain a concept or to make a point.

For example, when we presented our initial proposal to our agent, Elyse Cheney, she and her colleagues had trouble understanding deliberate practice clearly. In particular, they didn't get what separates deliberate practice from other forms of practice, other than that it is more effective. This was not their fault, but an indication that we hadn't made our explanation as readily intelligible as we'd thought. That forced us to rethink how we were presenting deliberate practice — in essence, to come up with a new and better mental representation of how we thought about it and how we wanted others to think about it. It soon occurred to us that the role of mental representations held the key to how we wanted to present deliberate practice.

Initially, we had seen mental representations as being just one aspect of deliberate practice among many that we would present to the reader, but now we began to see them as a central feature — perhaps *the* central feature — of the book. The main purpose of deliberate practice is to develop effective mental representations, and, as we will discuss shortly, mental representations in turn play a key role in deliberate practice. The key change that occurs in our adaptable brains in response to deliberate practice is the development of better mental representations, which in turn open up new possibilities for improved performance. In short, we came to see our explanation of mental representations as the keystone of the book, without which the rest of the book could not stand.

There was a steady interplay between the writing of the book and our conceptualization of the topic, and as we looked for ways to make our messages clearer to the reader, we would come up with new ways to think about deliberate practice ourselves. Researchers refer to this sort of writing as "knowledge transforming," as opposed to "knowledge telling," because the process of writing changes and adds to the knowledge that the writer had when starting out.

This is an example of one way in which expert performers use mental representations to improve their performance: they monitor and evaluate their performance, and, when necessary, they modify their mental representations in order to make them more effective. The more effective the mental representation is, the better the performance will be. We had developed a certain mental representation of the book, but we found out that it had led us to a performance (the explanations in our original proposal) that was not as good as we wished, so we used the feedback we had gotten and modified the representation accordingly. This in turn led us to a much better explanation of deliberate practice.

And so it went throughout the writing of the book. Although it was constantly evolving, our mental representation of the book guided us and informed our decisions about our writing. As we went along, we evaluated each piece — in the later stages with the help of our editor, Eamon Dolan — and when we found weaknesses, we tweaked the representation to fix the problem.

Obviously the mental representation for a book is much larger and more complex than one for a personal letter or a blog post, but the general pattern is the same: to write well, develop a mental representation ahead of time to guide your efforts, then monitor and evaluate your efforts and be ready to modify that representation as necessary.

MENTAL REPRESENTATIONS IN LEARNING

In general, mental representations aren't just the result of learning a skill; they can also help us learn. Some of the best evidence for this comes from the field of musical performance. Several researchers have examined what differentiates the best musicians from lesser ones, and one of the major differences lies in the quality of the mental representations the best ones create. When practicing a new piece, beginning and intermediate musicians generally lack a good, clear idea of how the music should sound, while advanced musicians have a very detailed mental representation of the music they use to guide their practice and, ultimately, their performance of a piece. In particular, they use their mental representations to provide their own feedback so that they know how close they are to getting the piece right and what they need to do differently to improve. The beginners and intermediate students may have crude representations of the music that allow them to tell, for instance, when they hit a wrong note, but they must rely on feedback from their teachers to identify the more subtle mistakes and weaknesses.

Even among beginning music students, it seems that differences in the quality of how the music is represented make a difference in how effective practice can be. About fifteen years ago two Australian psychologists, Gary McPherson and James Renwick, studied a number of children between the ages of seven and nine who were learning to play various instruments: the flute, the trumpet, the cornet, the clarinet, and the saxophone. Part of the study was to videotape the children as they practiced at home and then to analyze the practice sessions to understand what the children did to make their practice more or less effective.

In particular, the researchers counted the number of mistakes a student made in practicing a piece the first time and the second time and used the improvement from the first time to the second as a measurement of how effectively the student was practicing. They found a wide variation in the amount of improvement. Of all the students they studied, a female cornet player in her first year of learning the instrument made the most mistakes: 11 per minute, on average, on the first times playing pieces during practice sessions. On the second time through, she was still making the same mistakes 70 percent of the time — noticing and correcting only 3 out of every 10 mistakes. By contrast, the best first-year player, a boy who was learning the saxophone, made only 1.4 mistakes per minute on his first times through. And on the second times through, he was making the same mistakes only 20 percent of the time — correcting 8 out of every 10 mistakes. The difference in the percentage of corrections is particularly striking because the saxophone player was already making many fewer mistakes, so he had much less room for improvement.

All of the students had good attitudes and were motivated to improve, so McPherson and Renwick concluded that the differences among the students most likely lay, in large part, in how well the students were able to detect their mistakes — that is, how effective their mental representations of the musical pieces were. The saxophone player had a clear mental representation of the piece that allowed him to recognize most of his mistakes, remember them the next time, and correct them. The cornet player, on the other hand, didn't seem to have such a well-developed mental representation of what she was playing. The difference between the two was not in desire or effort, the researchers said. The cornet player just didn't have the same tools with which to improve as the saxophone player did.

McPherson and Renwick didn't try to understand the precise nature of the mental representations, but other research indicates that the representations could have taken several forms. One would be an aural representation — a clear idea of what a piece should sound like. Musicians at every level use these to guide their practice and their play, and better musicians have far more detailed representations, which include not just the pitch and the length of the notes to be played, but their volume, rise and fall, intonation, vibrato, tremolo, and harmonic relationship with other notes, including notes played on other instruments by other musicians. Good musicians not only recognize these various qualities of musical sound but know how to produce them on their instruments — an understanding that requires its own sort of mental representation, which is in turn quite closely tied to the mental representations of the sounds themselves.

The students McPherson and Renwick studied probably also had developed, to one degree or another, mental representations that connected notes written on a musical score with the fingering necessary to play those notes. Thus, if the saxophonist accidentally placed his fingers in the wrong position for a note, he would probably notice it not only because his horn produced the wrong sound but also because his fingering felt "off" — that is, it didn't match his mental representation of where his fingers should be placed.

While the study by McPherson and Renwick has the advantage of being very personal — we almost feel as though we know the cornet player and the saxophonist when we're done — it has the disadvantage of having observed only a few musicians in one school. Fortunately, its results are backed up by a British study of more than three thousand music students, ranging from beginners to experts ready to enter a university-level conservatory.

The researchers found, among other things, that the more accomplished music students were better able to determine when they'd made mistakes and better able to identify difficult sections they needed to focus their efforts on. This implies that the students had more highly developed mental representations of the music they were playing and of their own performances, which allowed them to monitor their practice and spot mistakes. Furthermore, the more advanced music students also had more effective practice techniques. The implication is that they were using their mental representations not only to spot mistakes but also to match appropriate practice techniques with the types of difficulties they were having with the music.

In any area, not just musical performance, the relationship between skill and mental representations is a virtuous circle: the more skilled you become, the better your mental representations are, and the better your mental representations are, the more effectively you can practice to hone your skill.

We can see a more detailed depiction of how an expert uses mental representations through a long-term collaboration between Roger Chaffin, a psychologist at the University of Connecticut, and Gabriela Imreh, an internationally known pianist based in New Jersey. For years they have been working together to understand what goes through Imreh's head as she studies, practices, and performs a piece of music.

Much of Chaffin's work with Imreh is reminiscent of how I monitored Steve Faloon's development of mental representations for memorizing strings of digits. He observes her as she is learning a new piece of music and has her voice her thought processes as she determines how she will play it. He also videotapes these practice sessions so that he has additional clues as to how Imreh is approaching her task.

In one series of sessions, Chaffin followed Imreh as she spent more than thirty hours practicing the third movement of the Italian Concerto of Johann Sebastian Bach, which she was scheduled to play for the first time. The first thing Imreh did when she sight-read the piece was to develop what he called an "artistic image"— a representation of what the piece should sound like when she performed it. Now, Imreh was not coming to this piece cold — she'd heard it many times — but the fact that she was able to create this mental image of the piece simply by reading the score indicates just how highly developed her mental representations of the piano are. Where most of us would see musical symbols on a page, she heard the music in her head.

Much of what Imreh did from that point on was figuring out how to perform the piece so that it matched her artistic image. She began by going through the entire piece and deciding exactly what fingering she would use. Where possible, she would use the standard fingering that pianists learn for particular series of notes, but there were places that required departing from the standard because she wanted that particular passage to sound a certain way. She would try out different options, decide on one, and note it on the score. She also identified different moments in the composition that Chaffin called "expressive turning points"—for instance, a point where her playing would turn from light and lively to more measured and serious. Later she would pick out cues in the music—short passages before a turning point or a technically difficult passage that, when she came to them, would serve as prompts to get ready for what was coming. She also picked out various places where she would add nuanced interpretations of the music.

By putting all of these different elements into an overall map of the piece, Imreh managed to do justice to both the forest and the trees. She formed an image of what the whole piece should sound like, while also giving herself clear images of the details she needed to pay close attention to as she was playing. Her mental representation combined what she thought the music was supposed to sound like with what Imreh had figured out about how to make it sound that way. Although other pianists' mental representations would likely differ from Imreh's in the specifics, their overall approaches are likely to be very similar.

Her mental representation also allowed Imreh to deal with a fundamental dilemma facing any classical pianist learning to play a piece. It is crucial that the musician practice and memorize the piece in such a way that the performance can be done almost automatically, with the fingers of each hand playing the proper notes with little or no conscious direction from the pianist; in this way the piece can be performed flawlessly on stage in front of an audience even if the pianist is nervous or excited. On the other hand, the pianist must have a certain amount of spontaneity in order to connect and communicate with the audience. Imreh did this by using her mental map of the piece. She would play much of the piece just as she always practiced it, with her fingers going through well-rehearsed motions, but she always knew exactly where she was in the piece because she'd identified various points that served as landmarks. Some of these were performance landmarks that would signal to Imreh that, for example, a change in fingering was approaching, while others were what Chaffin called "expressive landmarks." These indicated places where she could vary her playing to capture a particular emotion, depending on how she felt and how the audience was responding. That allowed her to maintain spontaneity within the demanding constraints of performing a complicated piece before a live audience.

PHYSICAL ACTIVITIES ARE MENTAL TOO

As we've just seen from several studies, musicians rely on mental representations to improve both the physical and cognitive aspects of their specialties. And mental representations are essential to activities we see as almost purely physical. Indeed, any expert in any field can be rightly seen as a high-achieving intellectual where that field is concerned. This applies to pretty much any activity in which the positioning and movement of a person's body is evaluated for artistic expression by human judges. Think of gymnastics, diving, figure skating, or dancing. Performers in these areas must develop clear mental representations of how their bodies are supposed to move to generate the artistic appearance of their performance routines. But even in areas where artistic form is not explicitly judged, it is still important to train the body to move in particularly efficient ways. Swimmers learn to perform their strokes in ways that maximize thrust and minimize drag. Runners learn to stride in ways that maximize speed and endurance while conserving energy. Pole-vaulters, tennis players, martial artists, golfers, hitters in baseball, three-point shooters in basketball, weightlifters, skeet shooters, and downhill skiers — for all of these athletes proper form is key to good performance, and the performers with the best mental representations will have an advantage over the rest.

In these areas too, the virtuous circle rules: honing the skill improves mental representation, and mental representation helps hone the skill. There is a bit of a chicken-and-egg component to this. Take figure skating: it's hard to have a clear mental representation of what a double axel feels like until you've done it, and, likewise, it is difficult to do a clean double axel without a good mental representation of one. That sounds paradoxical, but it isn't really. You work up to a double axel bit by bit, assembling the mental representations as you go.

It's like a staircase that you climb as you build it. Each step of your ascent puts you in a position to build the next step. Then you build that step, and you're in a position to build the next one. And so on. Your existing mental representations guide your performance and allow you to both monitor and judge that performance. As you push yourself to do something new — to develop a new skill or sharpen an old one — you are also expanding and sharpening your mental representations, which will in turn make it possible for you to do more than you could before.

4

The Gold Standard

WHAT IS MISSING from purposeful practice? What is required beyond simply focusing and pushing beyond one's comfort zone? Let's talk about it.

As we saw in chapter 1, purposeful practice as done by different people can have very different results. Steve Faloon reached the point where he could remember up to eighty-two digits, while Renée, working just as hard as Steve, was unable to get beyond twenty. The difference lay in the details of the types of practice that Steve and Renée used to improve their memory.

Since Steve first demonstrated that it is possible to memorize long strings of numbers, dozens of memory competitors have developed digit memories beyond what Steve achieved. According to the World Memory Sports Council, which oversees international memory competitions, there are now at least five people who have managed to remember 300 or more digits in a memory competition, and several dozen who have memorized at least 100 digits. As of November 2015 the world record in this event was held by Tsogbadrakh Saikhanbayar of Mongolia, who recalled 432 digits at the 2015 Taiwan Open Adult Memory Competition. That's more than five times as many digits as Steve's record. As with the disparity between Renée and Steve, the key difference between Steve's performance and that of the new generation of memory whizzes lies in the details of their training.

This is part of a general pattern. In every area, some approaches to training are more effective than others. In this chapter we'll explore the most effective method of all: deliberate practice. It is the gold standard, the ideal to which anyone learning a skill should aspire.

A HIGHLY DEVELOPED FIELD

Some activities, such as playing music in pop music groups, solving crossword puzzles, and folk dancing, have no standard training approaches. Whatever methods there are seem slapdash and produce unpredictable results. Other activities, like classical music performance, mathematics, and ballet, are blessed with highly developed, broadly accepted training methods. If one follows these methods carefully and diligently, one will almost surely become an expert. I've spent my career studying this second sort of field.

These fields have several characteristics in common. First, there are always objective ways — such as the win/loss of a chess competition or a head-to-head race — or at least semiobjective ways — such as evaluation by expert judges — to measure performance. This makes sense: if there is no agreement on what good performance is and no way to tell what changes would improve performance, then it is very difficult — often impossible — to develop effective training methods. If you don't know for sure what constitutes improvement, how can you develop methods to improve performance? Second, these fields tend to be competitive enough that performers have strong incentive to practice and improve. Third, these fields are generally well established, with the relevant skills having been developed over decades or even centuries. And fourth, these fields have a subset of performers who also serve as teachers and coaches and who, over time, have developed increasingly sophisticated sets of training techniques that make possible the field's steadily increasing skill level. The improvement of skills and the development of training techniques move forward hand in hand, with new training techniques leading to new levels of accomplishment and new accomplishments generating innovations in training. (The virtuous circle again.) This joint development of skills and training techniques has — up to now at least — always been carried out through trial and error, with a field's practitioners experimenting with various ways to improve, keeping what works and discarding what doesn't.

No field adheres more strongly to these principles than musical training, particularly on the violin and the piano. This is a competitive field and one in which the development of the requisite skills and training methods has been going on for several hundred years. Furthermore, it is an area that, at least in the case of the violin and piano, generally requires twenty or more years of steady practice if you are to take your place among the best in the world.

In short, it is a natural field — and quite likely the very best field — to study for anyone wishing to understand expert performance. And, luckily, it's the field I studied in the years after I had completed my research on expert performance in memory.

In the fall of 1987 I took a position at the Max Planck Institute for Human Development. After finishing my memory studies with Steve Faloon, I had followed up by studying other examples of exceptional memory, such as waiters who could recall the detailed orders of many customers without writing them down and stage actors who had to learn many lines every time they began a new play. In each case I had studied the mental representations that these people developed in order to build their memory, but they all had a major limitation: they were "amateurs" who had undergone no formal training but had just figured it out as they went along. What sorts of achievement might be possible with rigorous, formal training methods? When I moved to Berlin I suddenly had the chance to observe just such methods in musicians.

That opportunity arose thanks to the presence of the Universität der Künste Berlin — or, in English, the Berlin University of the Arts — which is located not far from the Max Planck Institute. The university has thirty-six hundred students in four colleges — a college of fine arts, a college of architecture, a college of media and design, and a college of music and the performing arts — and the music academy in particular is highly regarded for both its teaching and its student body. Its alumni include the conductors Otto Klemperer and Bruno Walter, two giants of twentieth-century conducting, and the composer Kurt Weill, best known for *The Threepenny Opera* and, in particular, for its popular song "Mack the Knife." Year after year the academy turns out pianists, violinists, composers, conductors, and other musicians who go on to take their places among Germany's — and the world's — elite artists.

At the Max Planck Institute, I recruited two collaborators — Ralf Krampe, a graduate student at the institute, and Clemens Tesch-Römer, a postdoctoral fellow there — and together we mapped out an investigation into the development of musical accomplishment. Originally the plan was to focus on the motivations of the music students. In particular, I was curious as to whether musicians' motivations would explain how much practice they engaged in — and thus explain at least in part how accomplished they became.

Ralf, Clemens, and I chose to limit ourselves to the academy's violin students. Because the school was well known for turning out world-class violinists, many of those students would likely rank among the world's best violinists in a decade or two. Not all of them were quite so accomplished, of course. The academy had a range of violin students, from good to very good to great, and this gave us the opportunity to compare the motivation of the various students with their levels of accomplishment.

We first asked the professors at the music academy to identify students who had the potential to have careers as international soloists — the very upper tier of professional violinists. These were the superstars-in-waiting, the students who intimidated all their classmates. The professors came up with fourteen names. Of those, three were not fluent in German — and thus would be difficult to interview — and one was pregnant and wouldn't be able to practice in her normal manner. That left us with ten "best" students — seven women and three men. The professors also identified a number of violin students who were very good but not superstar-good. We chose ten of them and matched them with the first ten by age and sex. These were the "better" students. Finally, we selected another ten age- and sex-matched violinists from the music-education department at the school. These students would most likely end up as music teachers, and while they were certainly skilled musicians when compared to the rest of us, they were clearly less skilled than the violinists in either of the other two groups. Many of the music teachers had unsuccessfully applied to be admitted to the soloist program and then had been accepted into the music-teacher program. This was our "good" group, which gave us three groups that had achieved very different levels of performance: good, better, and best.

We also recruited ten middle-aged violinists from the Berlin Philharmonic Orchestra (now the Berlin Philharmonic) and the Radio-Symphonie-Orchester Berlin, two orchestras with international reputations. The music teachers at the academy had told us that their best students were likely to end up performing in one of these orchestras or in ensembles of similar quality elsewhere in Germany; thus the violinists from these orchestras served as a look to the future — a glimpse of what the best violinists at the music academy were likely to be in another twenty or thirty years.

Our goal was to understand what separated the truly outstanding student violinists from those who were merely good. The traditional view held that differences among individuals performing at these highest levels would be due primarily to innate talent. So differences in the amount and type of practice — in essence, differences in motivation — wouldn't matter at this level. We were looking to see if this traditional view was wrong.

THE CHALLENGE OF THE VIOLIN

It is hard to describe the difficulty of playing a violin — and thus to explain how much skill a good violinist actually has — to someone whose only contact with the violin has been to hear it played by a professional. In the right hands no instrument sounds more beautiful, but put it in the wrong hands and you may as well step on a cat's tail and listen to the sounds that result. Coaxing just a single acceptable note from a violin — one that doesn't screech or squawk or whistle, one that is neither flat nor sharp, one that captures the tone of the instrument — requires a great deal of practice, and learning to play that single note well is just the first step in a long and challenging journey.

The difficulties start with the fact that the violin's fingerboard has no frets, the metal ridges found on a guitar's fingerboard that divide it into separate notes and guarantee that, as long as the guitar is in tune, each note played will sound neither flat nor sharp. Because the violin has no frets, the violinist must put his or her fingers at exactly the right spot on the fingerboard to produce the desired note. A sixteenth of an inch off the mark, and the note will be flat or sharp. If the finger is too far from the correct position, the result is a completely different note from the one that was desired. And that's just one note; every note up and down the fingerboard requires the same precision. Violinists spend countless hours doing scales so that they can move the fingers of their left hand correctly from one note to the next, whether up or down on a single string or moving from one string to another. And once they are comfortable with placing their fingers in exactly the right spots on the fingerboard, there are various subtleties of fingering to master, beginning with vibrato, which is a rolling — not a sliding — of the fingertip up and down the string, which causes the note to shimmer. More hours and hours of practice.

Furthermore, the fingering is actually the easy part. Using the bow properly poses another whole level of difficulty. As the bow is drawn across a string, the horsehair of the bow catches the string and drags it a bit, then lets it slip, catches it again, lets it slip, and so on hundreds or even thousands of times a second, depending on the frequency of the string's vibrations. The particular way that the string moves in response to the bow's drag-and-release action gives the violin its distinctive sound. Violinists control the volume of their playing by varying the pressure of the bow on the string, but that pressure must stay within a certain range; too much and the result is an awful squawking noise, while too little leads to a sound that, while less offensive, isn't considered acceptable. To complicate matters further, the range of acceptable pressures varies according to the bow's position along the string. The closer the bow is to the bridge, the more force is needed to stay within the sweet spot.

Violinists must learn to move the bow across the strings in a variety of ways in order to vary the sound that is produced. The bow can be drawn smoothly across the strings, stopped momentarily, sawed quickly back and forth, picked up and dropped back down on the strings, allowed to bounce gently off the strings, and so on — more than a dozen bowing techniques in all. *Spiccato,* for example, involves bouncing the bow off and back onto a string as the bow moves back and forth across the string, producing a series of short, staccato notes. *Sautillé* is a faster version of *spiccato.* Then there are *jeté, collé, détaché, martelé, legato, louré,* and more, each technique with its own distinctive sound. And, of course, all of these bow techniques must be done in close coordination with the left hand as it fingers the strings.

These are not skills that can be picked up in a year or two of practice. Indeed, all of the students we studied had been playing for well over a decade — the average age at which they started was eight — and they had all followed the training pattern that is standard for children today. That is, they began systematic, focused lessons very early on, visiting a music teacher usually once a week. During that weekly meeting, the student's current musical performance was evaluated by the teacher, who identified a couple of immediate goals for improvement and assigned some practice activities that a motivated student would be able to attain with solitary practice during the week before the next meeting.

Because most students spend the same amount of time each week with the music teacher — an hour — the primary difference in practice from one student to the next lies in how much time the students devote to solitary practice. Among serious students — such as the ones who ended up in the Berlin academy — it's not unusual for ten- and eleven-year-olds to be spending fifteen hours a week on focused practice, during which time they are following lessons designed by their teachers to develop specific techniques. And as they get older, the serious students generally increase their amount of weekly practice time.

One of the things that differentiates violin training from training in other areas — soccer, for example, or algebra — is that the set of skills expected of a violinist is quite standardized, as are many of the instruction techniques. Because most violin techniques are decades or even centuries old, the field has had the chance to zero in on the proper or "best" way to hold the violin, to move the hand during vibrato, to move the bow during *spiccato,* and so on. The various techniques may not be easy to master, but a student can be shown exactly what to do and how to do it.

All this means is that the violin students at the Universität der Künste Berlin offered a near-perfect opportunity to test the role that motivation plays in developing expert performance and, more generally, to identify what differentiates good performers from the very best.

GOOD VERSUS BETTER VERSUS BEST

To look for these differences, we interviewed each of the thirty student violinists in our study in great detail. We asked them about their musical histories — when they started studying music, who their teachers were, how many hours a week they spent in solitary practice at each age, what competitions they'd won, and so on. We asked them for their opinions on how important various activities were in improving their performance — practicing alone, practicing in a group, playing alone for fun, playing in a group for fun, performing solo, performing in a group, taking lessons, giving lessons, listening to music, studying music theory, and so on. We asked them how much effort these various activities required and how much immediate pleasure they got while they were doing them. We asked them to estimate how much time they'd spent on each of these activities during the previous week. Finally, because we were interested in how much time they'd spent on practice over the years, we asked them to estimate, for each year since they'd started to practice music, how many hours per week on average they had spent in solitary practice.

The thirty music students were also asked to keep daily diaries for each of the next seven days in which they would detail exactly how they'd spent their time. In the diaries, they recorded their activities in fifteen-minute increments: sleeping, eating, going to class, studying, practicing alone, practicing with others, performing, and so on. When they were done we had a detailed picture of how they'd spent their days as well as a very good idea of their practice histories.

The students from all three groups gave similar answers to most of our questions. The students pretty much all agreed, for instance, that solitary practice was the most important factor in improving their performance, followed by such things as practicing with others, taking lessons, performing (particularly in solo performance), listening to music, and studying music theory. Many of them also said that getting enough sleep was very important to their improvement. Because their practice was so intense, they needed to recharge their batteries with a full night's sleep — and often an afternoon nap.

One of our most significant findings was that most factors the students had identified as being important to improvement were also seen as labor-intensive and not much fun; the only exceptions were listening to music and sleeping. Everyone from the very top students to the future music teachers agreed: improvement was hard, and they didn't enjoy the work they did to improve. In short, there were no students who just loved to practice and thus needed less motivation than the others. These students were motivated to practice intensely and with full concentration because they saw such practice as essential to improving their performance.

The other crucial finding was that there was only one major difference among the three groups. This was the total number of hours that the students had devoted to solitary practice.

Using the students' estimates of how many hours a week they'd practiced alone since they'd begun playing the violin, we calculated the total number of hours they'd spent practicing alone until age eighteen, the age at which they typically entered the music academy. Although memories are not always reliable, dedicated students of this sort generally set aside fixed periods to practice each day on a weekly schedule — and they do this beginning very early on in their music training — so we thought it likely that their retrospective estimates of how much time they had spent practicing at various ages would be relatively accurate.

We found that the best violin students had, on average, spent significantly more time than the better violin students had spent, and that the top two groups — better and best — had spent much more time on solitary practice than the music-education students. Specifically, the music-education students had practiced an average of 3,420 hours on the violin by the time they were eighteen, the better violin students had practiced an average of 5,301 hours, and the best violin students had practiced an average of 7,410 hours. Nobody had been slacking—even the least accomplished of the students had put in thousands of hours of practice, far more than anyone would have who played the violin just for fun—but these were clearly major differences in practice time.

Looking more closely, we found that the largest differences in practice time among the three groups of students had come in the preteen and teenage years. This is a particularly challenging time for young people to keep up their music practice because of the many interests that compete for their time—studying, shopping, hanging out with friends, partying, and so on. Our results indicated that those preteens and teens who could maintain and even increase their heavy practice schedule during these years ended up in the top group of violinists at the academy.

We also calculated estimated practice times for the middle-aged violinists working at the Berlin Philharmonic and the Radio-Symphonie-Orchester Berlin, and we found that the time they had spent practicing before the age of eighteen—an average of 7,336 hours—was almost identical to what the best violin students in the music academy had reported.

There were a number of factors we did not include in our study that could have influenced—and indeed probably *did* influence—the skill levels of the violinists in the different groups. For instance, students who were lucky enough to have worked with exceptional teachers would likely have progressed more quickly than those with teachers who were just okay.

But two things were strikingly clear from the study: First, to become an excellent violinist requires several thousand hours of practice. We found no shortcuts and no "prodigies" who reached an expert level with relatively little practice. And, second, even among these gifted musicians—all of whom had been admitted to the best music academy in Germany—the violinists who had spent significantly more hours practicing their craft were on average more accomplished than those who had spent less time practicing.

The same pattern that we saw among the student violinists has been seen among performers in other areas. Observing this pattern accurately depends on being able to get a good estimate of the total number of hours of practice people have put into developing a skill — which is not always easy to do — and also on being able to tell with some objectivity who the good, better, and best are in a given field, which is also not always easy to do. But when you can do those two things, you generally find that the best performers are those who have spent the most time in various types of purposeful practice.

Just a few years ago I and two colleagues, Carla Hutchinson and Natalie Sachs-Ericsson (who is also my wife), studied a group of ballet dancers to see what role practice played in their achievements. The dancers we worked with were from the Bolshoi Ballet in Russia, the National Ballet of Mexico, and three companies in the United States: the Boston Ballet, the Dance Theatre of Harlem, and the Cleveland Ballet. We gave them questionnaires to learn when they started training and how many hours a week they devoted over time to practice — which consisted mainly of practice time spent in a studio under the direction of an instructor — and we specifically excluded rehearsals and performances. We judged a dancer's skill level by determining what sort of ballet company he or she had performed with — a regional company, such as the Cleveland Ballet, or a national company, such as the Dance Theatre of Harlem, or an international company, such as the Bolshoi or the Boston Ballet — and also by determining the highest level the dancer had reached inside the company, whether a principal dancer, a soloist, or just a member of the troupe. The average age of the dancers was twenty-six, but the youngest was eighteen, so to have an apples-to-apples comparison, we looked at the accumulated amount of practice through age seventeen and the skill level at age eighteen.

Though we were working with fairly crude measures — both of the total hours of practice and of the dancers' abilities — there was still a relatively strong relationship between the reported amount of time spent on practice and how high a dancer had risen in the world of ballet, with the dancers who practiced more being better dancers, at least according to the troupes they danced with and the positions they held in the troupes. There was no significant difference between dancers from different countries in terms of how many hours of practice they needed to reach a certain level of proficiency.

As with the violinists, the only significant factor determining an individual ballet dancer's ultimate skill level was the total number of hours devoted to practice. When we calculated how much time the dancers had spent on practice through age twenty, we found that they had averaged more than ten thousand hours of practice. Some dancers had put in much more time than this average, however, while others had put in much less, and this difference in training corresponded to the difference between good, better, and best among the dancers. Again, we found no sign of anyone born with the sort of talent that would make it possible to reach the upper levels of ballet without working as hard or harder than anyone else. Other studies of ballet dancers have shown the same thing.

By now it is safe to conclude from many studies on a wide variety of disciplines that nobody develops extraordinary abilities without putting in tremendous amounts of practice. I do not know of any serious scientist who doubts that conclusion. No matter which area you study — music, dance, sports, competitive games, or anything else with objective measures of performance — you find that the top performers have devoted a tremendous amount of time to developing their abilities. We know from studies of the world's best chess players, for example, that almost no one reaches the level of grandmaster with less than a decade of intense study. Even Bobby Fischer, who at the time was the youngest person ever to become a grandmaster and whom many consider to have been the greatest chess player in history, studied chess for nine years before he reached grandmaster level. Since Fischer's achievement, others have achieved grandmaster status at increasingly younger ages, as advances in training and practice methods have made it possible for young players to improve ever more quickly, but it still takes many years of sustained practice to become a grandmaster.

THE PRINCIPLES OF DELIBERATE PRACTICE

In the most highly developed fields — the ones that have benefited from many decades or even centuries of steady improvement, with each generation passing on the lessons and skills it has learned to the next — the approach to individualized practice is amazingly uniform. No matter where you look — musical performance, ballet, or sports such as figure skating or gymnastics — you will find that training follows a very similar set of principles. That study of the Berlin violin students introduced me to this sort of practice, which I named "deliberate practice," and I have since studied it in many other fields. When my colleagues and I published our results on the violin students, we described deliberate practice as follows.

We began by noting that the levels of performance in such areas as musical performance and sports activities have increased greatly over time, and that as individuals have developed greater and more complex skills and performance, teachers and coaches have developed various methods to teach these skills. The improvement in performance generally has gone hand in hand with the development of teaching methods, and today anyone who wishes to become an expert in these fields will need an instructor's help. Because few students can afford a full-time teacher, the standard pattern is to have a lesson once or a few times in a week, with the teachers assigning practice activities the student is expected to perform between lessons. These activities are generally designed with the student's current abilities in mind and are intended to push him or her to move just beyond the current skill level. It was these practice activities that my colleagues and I defined as "deliberate practice."

In short, we were saying that deliberate practice is different from other sorts of purposeful practice in two important ways: First, it requires a field that is already reasonably well developed — that is, a field in which the best performers have attained a level of performance that clearly sets them apart from people who are just entering the field. We're referring to activities like musical performance (obviously), ballet and other sorts of dance, chess, and many individual and team sports, particularly the sports in which athletes are scored for their individual performance, such as gymnastics, figure skating, or diving. What areas don't qualify? Pretty much anything in which there is little or no direct competition, such as gardening and other hobbies, for instance, and many of the jobs in today's workplace — business manager, teacher, electrician, engineer, consultant, and so on. These are not areas where you're likely to find accumulated knowledge about deliberate practice, simply because there are no objective criteria for superior performance.

Second, deliberate practice requires a teacher who can provide practice activities designed to help a student improve his or her performance. Of course, before there can be such teachers there must be individuals who have achieved a certain level of performance with practice methods that can be passed on to others.

With this definition we are drawing a clear distinction between purposeful practice — in which a person tries very hard to push himself or herself to improve — and practice that is both purposeful and *informed*. In particular, deliberate practice is informed and guided by the best performers' accomplishments and by an understanding of what these expert performers do to excel. Deliberate practice is purposeful practice that knows where it is going and how to get there.

In short, deliberate practice is characterized by the following traits:

- Deliberate practice develops skills that other people have already figured out how to do and for which effective training techniques have been established. The practice regimen should be designed and overseen by a teacher or coach who is familiar with the abilities of expert performers and with how those abilities can best be developed.
- Deliberate practice takes place outside one's comfort zone and requires a student to constantly try things that are just beyond his or her current abilities. Thus it demands near-maximal effort, which is generally not enjoyable.
- Deliberate practice involves well-defined, specific goals and often involves improving some aspect of the target performance; it is not aimed at some vague overall improvement. Once an overall goal has been set, a teacher or coach will develop a plan for making a series of small changes that will add up to the desired larger change. Improving some aspect of the target performance allows a performer to see that his or her performances have been improved by the training.
- Deliberate practice is deliberate, that is, it requires a person's full attention and conscious actions. It isn't enough to simply follow a teacher's or coach's directions. The student must concentrate on the specific goal for his or her practice activity so that adjustments can be made to control practice.
- Deliberate practice involves feedback and modification of efforts in response to that feedback. Early in the training process much of the feedback will come from the teacher or coach, who will monitor progress, point out problems, and offer ways to address those problems. With time and experience students must learn to monitor themselves, spot mistakes, and adjust accordingly. Such self-monitoring requires effective mental representations.
- Deliberate practice both produces and depends on effective mental representations. Improving performance goes hand in hand

with improving mental representations; as one's performance improves, the representations become more detailed and effective, in turn making it possible to improve even more. Mental representations make it possible to monitor how one is doing, both in practice and in actual performance. They show the right way to do something and allow one to notice when doing something wrong and to correct it.

- Deliberate practice nearly always involves building or modifying previously acquired skills by focusing on particular aspects of those skills and working to improve them specifically; over time this step-by-step improvement will eventually lead to expert performance. Because of the way that new skills are built on top of existing skills, it is important for teachers to provide beginners with the correct fundamental skills in order to minimize the chances that the student will have to relearn those fundamental skills later when at a more advanced level.

APPLYING THE PRINCIPLES OF DELIBERATE PRACTICE

As defined, deliberate practice is a very specialized form of practice. You need a teacher or coach who assigns practice techniques designed to help you improve on very specific skills. That teacher or coach must draw from a highly developed body of knowledge about the best way to teach these skills. And the field itself must have a highly developed set of skills that are available to be taught. There are relatively few fields — musical performance, chess, ballet, gymnastics, and the rest of the usual suspects — in which all of these things are true and it is possible to engage in deliberate practice in the strictest sense.

But not to worry — even if your field is one in which deliberate practice in the strictest sense is not possible, you can still use the principles of deliberate practice as a guide to developing the most effective sort of practice possible in your area.

For a simple example, let's return once more to memorizing strings of digits. When Steve was working to improve the number of digits he could remember, he was obviously not using deliberate practice to improve. At the time there was no one who could remember forty or fifty digits, and there were records of only a handful of mnemonists who could remember more than fifteen. There were no known training methods, and, naturally, there were no teachers offering lessons. Steve had to figure it out through trial and error.

Today, many people — hundreds or more — train to remember digit strings in order to take part in memory competitions. Some people can recall three hundred and more digits. How do they do it? Not through deliberate practice, at least in its strictest sense. As far as I know, there are no digit-memory instructors out there.

However, something is different today than it was when Steve Faloon was practicing: there are now some well-known techniques for training your memory for long strings of digits. These techniques tend to be variants of the method that Steve developed — that is, they rely on memorizing chunks of two or three or four digits and then arranging those groups in a retrieval structure so that they can be recalled in order later.

I saw such a technique in action when I worked with Yi Hu to study one of the best digit memorizers in the world, Feng Wang of China. At the 2011 World Memory Championships, Feng set what was then the world record by recalling three hundred digits spoken at one per second. Once Professor Hu's assistant had tested Feng's memory encoding technique, it was clear to me that his method was similar to Steve's in spirit but quite different — and much more carefully designed — in its details. Feng based his methods on some of the well-known techniques I mentioned above.

Feng started by developing a set of memorable images that he associated with each of the hundred pairs of digits from 00 through 99. Next he developed a "map" of physical locations that he could visit in his mind in a very specific order. This is a latter-day version of "the memory palace" that people have used since the time of the ancient Greeks to remember large amounts of information. When Feng hears a string of digits, he takes each set of four numbers, encodes it as a pair of images corresponding to the first two digits in the set and the second two, and mentally places that pair of images in the appropriate location along his mental map. For example, in one trial he encoded the four-digit string 6389 as a banana (63) and a monk (89) and then mentally placed them in a pot; to remember the image, he thought to himself, "There is a banana in the pot, a monk split the banana." Once all of the digits in the list have been read out, Feng recalls the numbers by mentally traveling along the route of his map, remembering which images sat in each location, and then translating those images back into the corresponding numbers. Like Steve before him, Feng is enlisting his long-term memory, creating associations between the numbers in the string and items already in his long-term memory, thus moving far beyond the limitations imposed by short-term memory. But Feng is doing it in a much more sophisticated and effective way than Steve was.

Today's memory competitors can learn from the experiences of those who came before them. They identify the best practitioners — an easy task because it comes down to who can memorize the most digits — and then they determine what enabled these practitioners to perform so well and develop training techniques that will produce those same abilities themselves. While they may lack teachers to design their practice sessions, they can draw on the advice previous experts have recorded in books or interviews. And memory experts will often help others who want to acquire similar skills. Thus, while digit-memory training isn't deliberate practice in its strictest sense, it captures the most important element — learning from the best predecessors — and that has proved enough to generate rapid improvements in the field.

This is the basic blueprint for getting better in any pursuit: get as close to deliberate practice as you can. If you're in a field where deliberate practice is an option, you should take that option. If not, apply the principles of deliberate practice as much as possible. In practice this often boils down to purposeful practice with a few extra steps: first, identify the expert performers, then figure out what they do that makes them so good, then come up with training techniques that allow you to do it, too.

In determining who the experts are, the ideal is to use some objective measure to separate the best from the rest. This is relatively easy in those areas that involve direct competition, such as individual sports and games. It is also reasonably straightforward to pick out the best performers in the performing arts, which, while more dependent on subjective judgments, still involves well-accepted standards for performance and clear expectations for what expert performers do. (When athletes or performers are part of a group, it becomes trickier, but still there are often clear ideas about which individuals are among the best, the middle, or the weakest part of the group.) In other areas, however, it can be quite difficult to identify the true experts. How does one identify, for example, the best doctors or the best pilots or the best teachers? What does it even mean to speak of the best business managers or the best architects or the best advertising executives?

If you are trying to identify the best performers in an area that lacks rules-based, head-to-head competition or clear, objective measures of performance (such as scores or times), keep this one thing at the front of your mind: subjective judgments are inherently vulnerable to all sorts of biases. Research has shown that people are swayed by factors like education, experience, recognition, seniority, and even friendliness and attractiveness when they are judging another person's overall competence and expertise. We have already noted, for instance, how people often assume that more experienced doctors are better than less experienced ones, and people also assume that someone with several degrees will be more competent than someone with one or none. Even in the judgment of musical performance, which should be more objective than in most fields, research has shown that judges can be influenced by such irrelevant factors as the performer's reputation, sex, and physical attractiveness.

In many fields, people who are widely accepted as "experts" are actually not expert performers when judged by objective criteria. One of my favorite examples of this phenomenon concerns wine "experts." Many of us assume that their highly developed palates can pick out subtleties and nuances in wines that are not apparent to the rest of us, but studies have shown that their powers are highly exaggerated. For example, while it has long been known that the ratings given to individual wines often vary widely from expert to expert, a 2008 article in the *Journal of Wine Economics* reported that wine experts don't even agree with themselves.

Robert Hodgson, the owner of a small California winery, got in touch with the head judge of the annual wine competition at the California State Fair, in which thousands of wines are entered each year, and suggested an experiment. The competition is set up so that each judge tastes a flight of thirty wines at a time. The wines are not identified, so the judge cannot be influenced by reputation or other factors. Hodgson suggested that in a number of those flights, the judges should be given three samples of the same wine. Would they give these identical samples the same rating, or would their ratings vary?

The head judge agreed, and Hodgson ran this experiment at four consecutive state fairs from 2005 to 2008. He found that very few judges rated the three identical samples similarly. It was common for a judge to give scores that varied by plus or minus four points — that is, to give one sample a 91, a second sample of the same wine an 87, and the third an 83. This is a significant difference: a 91 wine is a good wine that will fetch a premium price, while an 83 is nothing special. Some judges determined one of the three samples to be worthy of a gold medal and another of the three to be worth just a bronze medal — or no medal at all. And while in any given year some judges were more consistent than others, when Hodgson compared them year to year, he found that judges who were consistent one year were inconsistent the next. None of the judges — and these were sommeliers, wine critics, winemakers, wine consultants, and wine buyers — proved to be consistent all the time.

Research has shown that the "experts" in many fields don't perform reliably better than other, less highly regarded members of the profession — or sometimes even than people who have had no training at all. In his influential book *House of Cards: Psychology and Psychotherapy Built on Myth,* the psychologist Robyn Dawes described research showing that licensed psychiatrists and psychologists were no more effective at performing therapy than laypeople who had received minimal training. Similarly, many studies have found that the performance of financial "experts" in picking stocks is little or no better than the performance of novices or random chance. And, as we noted earlier, doctors in general practice with several decades of experience sometimes perform worse, when judged by objective measures, than doctors with just a few years of experience — mainly because the younger doctors attended medical school more recently, so their training is more up-to-date and they are more likely to remember it. Contrary to expectations, experience doesn't lead to improved performance among many types of doctors and nurses.

The lesson here is clear: be careful when identifying expert performers. Ideally you want some objective measure of performance with which to compare people's abilities. If no such measures exist, get as close as you can. For example, in areas where a person's performance or product can be observed directly — a screenwriter, say, or a programmer — the judgment of peers is a good place to start, while keeping in mind the possible influence of unconscious bias. However, many professionals, including doctors, psychotherapists, and teachers, work mostly by themselves, and other professionals in their field may know little about their practices or about their outcomes with patients and students. Thus a good rule of thumb is to seek out people who work intimately with many other professionals, such as a nurse who plays a role on several different surgery teams and can compare their performance and identify the best. Another method is to seek out the persons that professionals themselves seek out when they need help with a particularly difficult situation. Talk to the people about who they think are the best performers in their field, but be certain that you ask them what type of experience and knowledge they have to be able to judge one professional as being better than another.

In a field you're already familiar with — like your own job — think carefully about what characterizes good performance and try to come up with ways to measure that, even if there must be a certain amount of subjectivity in your measurement. Then look for those people who score highest in the areas you believe are key to superior performance. Remember that the ideal is to find objective, reproducible measures that consistently distinguish the best from the rest, and if that ideal is not possible, approximate it as well as you can.

Once you've identified the expert performers in a field, the next step is to figure out specifically what they do that separates them from other, less accomplished people in the same field, and what training methods helped them get there. This is not always easy. Why does one teacher improve students' performances more than another? Why does one surgeon have better outcomes than another? Why does one salesperson consistently make more sales than another? You can generally bring an expert in the field in to observe the performance of various individuals and make suggestions about what they are doing well and what they need to improve on, but it may not be obvious, even to experts, exactly what differentiates the best performers from everyone else.

Part of the problem is the key role that mental representations play. In many fields it is the quality of mental representations that sets apart the best from the rest, and mental representations are, by their nature, not directly observable. Consider once more the task of memorizing strings of digits. Someone who watched a film of Steve Faloon repeating back a string of eighty-two digits and then saw Feng Wang doing three hundred would obviously know who was better, but there would be no way to know why. I myself know why, because having spent two years collecting verbal reports on Steve's thought processes and designing experiments to test ideas about his mental representations, I was able to use the same methods when my colleague Yi Hu and I studied Feng Wang. Having studied half a dozen memory experts' mental representations made it easier for me to identify the critical differences between Steve and Feng, but this is the exception rather than the rule. Even psychology researchers are only now just beginning to explore the role of mental representations in understanding why some people perform so much better than others, and there are very few areas in which we can say with certainty, "Here are the types of mental representations that the expert performers in the field use, and this is why they are more effective than other sorts of mental representations that one might use." If you have a psychological bent, it may be worthwhile to talk to the expert performers and try to get a sense of how they approach tasks and why. Even with that approach, however, you're likely to uncover just a small part of what makes them special, for often even they don't know. We'll discuss more about this in chapter 7.

Fortunately, in some cases you can bypass figuring out what sets experts themselves apart from others and simply figure out what sets their training apart. For instance, in the 1920s and 1930s the Finnish runner Paavo Nurmi set twenty-two world records in distances from 1,500 meters (just under a mile) to 20 kilometers (just under 12.5 miles). For a few years he was untouchable at any distance he chose to train for; everyone else was competing for second place. But eventually other runners realized that Nurmi's advantage came from having developed new training techniques, such as pacing himself with a stopwatch, using interval training to build speed, and following a year-long training regimen so that he was always training. Once those techniques were widely adopted, it elevated the performance of the entire field.

Lesson: Once you have identified an expert, identify what this person does differently from others that could explain the superior performance. There are likely to be many things the person does differently that have nothing to do with the superior performance, but at least it is a place to start.

In all of this keep in mind that the idea is to inform your purposeful practice and point it in directions that will be more effective. If you find that something works, keep doing it; if it doesn't work, stop. The better you are able to tailor your training to mirror the best performers in your field, the more effective your training is likely to be.

And finally remember that, whenever possible, the best approach is almost always to work with a good coach or teacher. An effective instructor will understand what must go into a successful training regimen and will be able to modify it as necessary to suit individual students.

Working with such a teacher is particularly important in areas like musical performance or ballet, where it takes ten-plus years to become an expert and where the training is cumulative, with the successful performance of one skill often depending on having previously mastered other skills. A knowledgeable instructor can lead the student to develop a good foundation and then gradually build on that foundation to create the skills expected in that field. In learning the piano, for instance, a student must have proper finger placement from the start, for while it may be possible to play simpler pieces with the fingers not in their ideal positions, more complicated pieces will demand that the student have developed proper habits. An experienced teacher will understand this; no student, no matter how motivated, can expect to figure out such things on his or her own.

Finally, a good teacher can give you valuable feedback you couldn't get any other way. Effective feedback is about more than whether you did something right or wrong. A good math teacher, for instance, will look at more than the answer to a problem; he'll look at exactly how the student got the answer as a way of understanding the mental representations the student was using. If needed, he'll offer advice on how to think more effectively about the problem.

NO, THE TEN-THOUSAND-HOUR RULE ISN'T REALLY A RULE

Ralf Krampe, Clemens Tesch-Römer, and I published the results from our study of the Berlin violin students in 1993. These findings would go on to become a major part of the scientific literature on expert performers, and over the years a great many other researchers have referred to them. But it was actually not until 2008, with the publication of Malcolm Gladwell's *Outliers,* that our results attracted much attention from outside the scientific community. In his discussion of what it takes to become a top performer in a given field, Gladwell offered a catchy phrase: "the ten-thousand-hour rule." According to this rule, it takes ten thousand hours of practice to become a master in most fields. We had indeed mentioned this figure in our report as the average number of hours that the best violinists had spent on solitary practice by the time they were twenty. Gladwell himself estimated that the Beatles had put in about ten thousand hours of practice while playing in Hamburg in the early 1960s and that Bill Gates put in roughly ten thousand hours of programming to develop his skills to a degree that allowed him to found and develop Microsoft. In general, Gladwell suggested, the same thing is true in essentially every field of human endeavor — people don't become expert at something until they've put in about ten thousand hours of practice.

The rule is irresistibly appealing. It's easy to remember, for one thing. It would've been far less effective if those violinists had put in, say, eleven thousand hours of practice by the time they were twenty. And it satisfies the human desire to discover a simple cause-and-effect relationship: just put in ten thousand hours of practice at anything, and you will become a master.

Unfortunately, this rule — which is the only thing that many people today know about the effects of practice — is wrong in several ways. (It is also correct in one important way, which I will get to shortly.) First, there is nothing special or magical about ten thousand hours. Gladwell could just as easily have mentioned the average amount of time the best violin students had practiced by the time they were eighteen — approximately seventy-four hundred hours — but he chose to refer to the total practice time they had accumulated by the time they were twenty, because it was a nice round number. And, either way, at eighteen or twenty, these students were nowhere near masters of the violin. They were very good, promising students who were likely headed to the top of their field, but they still had a long way to go when I studied them. Pianists who win international piano competitions tend to do so when they're around thirty years old, and thus they've probably put in about twenty thousand to twenty-five thousand hours of practice by then; ten thousand hours is only halfway down that path.

And the number varies from field to field. Steve Faloon became the very best person in the world at memorizing strings of digits after only about two hundred hours of practice. I don't know exactly how many hours of practice the best digit memorizers put in today before they get to the top, but it is likely well under ten thousand.

Second, the number of ten thousand hours at age twenty for the best violinists was only an average. Half of the ten violinists in that group hadn't actually accumulated ten thousand hours at that age. Gladwell misunderstood this fact and incorrectly claimed that *all* the violinists in that group had accumulated over ten thousand hours.

Third, Gladwell didn't distinguish between the deliberate practice that the musicians in our study did and any sort of activity that might be labeled "practice." For example, one of his key examples of the ten-thousand-hour rule was the Beatles' exhausting schedule of performances in Hamburg between 1960 and 1964. According to Gladwell, they played some twelve hundred times, each performance lasting as much as eight hours, which would have summed up to nearly ten thousand hours. *Tune In,* an exhaustive 2013 biography of the Beatles by Mark Lewisohn, calls this estimate into question and, after an extensive analysis, suggests that a more accurate total number is about eleven hundred hours of playing. So the Beatles became worldwide successes with far less than ten thousand hours of practice. More importantly, however, performing isn't the same thing as practice. Yes, the Beatles almost certainly improved as a band after their many hours of playing in Hamburg, particularly because they tended to play the same songs night after night, which gave them the opportunity to get feedback — both from the crowd and themselves — on their performance and find ways to improve it. But an hour of playing in front of a crowd, where the focus is on delivering the best possible performance at the time, is not the same as an hour of focused, goal-driven practice that is designed to address certain weaknesses and make certain improvements — the sort of practice that was the key factor in explaining the abilities of the Berlin student violinists.

A closely related issue is that, as Lewisohn argues, the success of the Beatles was not due to how well they performed other people's music but rather to their songwriting and creation of their own new music. Thus, if we are to explain the Beatles' success in terms of practice, we need to identify the activities that allowed John Lennon and Paul McCartney — the group's two primary songwriters — to develop and improve their skill at writing songs. All of the hours that the Beatles spent playing concerts in Hamburg would have done little, if anything, to help Lennon and McCartney become better songwriters, so we need to look elsewhere to explain the Beatles' success.

This distinction between deliberate practice aimed at a particular goal and generic practice is crucial because not every type of practice leads to the improved ability that we saw in the music students or the ballet dancers. Generally speaking, deliberate practice and related types of practice that are designed to achieve a certain goal consist of individualized training activities — usually done alone — that are devised specifically to improve particular aspects of performance.

The final problem with the ten-thousand-hour rule is that, although Gladwell himself didn't say this, many people have interpreted it as a promise that almost anyone can become an expert in a given field by putting in ten thousand hours of practice. But nothing in my study implied this. To show a result like this, I would have needed to put a collection of randomly chosen people through ten thousand hours of deliberate practice on the violin and then see how they turned out. All that our study had shown was that among the students who had become good enough to be admitted to the Berlin music academy, the best students had put in, on average, significantly more hours of solitary practice than the better students, and the better and best students had put in more solitary practice than the music-education students.

The question of whether anyone can become an expert performer in a given field by taking part in enough designed practice is still open, and I will offer some thoughts on this issue in the next chapter. But there was nothing in the original study to suggest that it was so.

Gladwell did get one thing right, and it is worth repeating because it's crucial: becoming accomplished in any field in which there is a well-established history of people working to become experts requires a tremendous amount of effort exerted over many years. It may not require exactly ten thousand hours, but it will take a lot.

We have seen this in chess and the violin, but research has shown something similar in field after field. Authors and poets have usually been writing for more than a decade before they produce their best work, and it is generally a decade or more between a scientist's first publication and his or her most important publication — and this is in addition to the years of study before that first published research. A study of musical composers by the psychologist John R. Hayes found that it takes an average of twenty years from the time a person starts studying music until he or she composes a truly excellent piece of music, and it is generally never less than ten years. Gladwell's ten-thousand-hour rule captures this fundamental truth — that in many areas of human endeavor it takes many, many years of practice to become one of the best in the world—in a forceful, memorable way, and that's a good thing.

On the other hand, emphasizing what it takes to become one of the best in the world in such competitive fields as music, chess, or academic research leads us to overlook what I believe to be the more important lesson from our study of the violin students. When we say that it takes ten thousand—or however many—hours to become really good at something, we put the focus on the daunting nature of the task. While some may take this as a challenge—as if to say, "All I have to do is spend ten thousand hours working on this, and I'll be one of the best in the world!"—many will see it as a stop sign: "Why should I even try if it's going to take me ten thousand hours to get really good?" As Dogbert observed in one *Dilbert* comic strip, "I would think a willingness to practice the same thing for ten thousand hours is a mental disorder."

But I see the core message as something else altogether: In pretty much any area of human endeavor, people have a tremendous capacity to improve their performance, as long as they train in the right way. If you practice something for a few hundred hours, you will almost certainly see great improvement—think of what two hundred hours of practice brought Steve Faloon—but you have only scratched the surface. You can keep going and going and going, getting better and better and better. How much you improve is up to you.

This puts the ten-thousand-hour rule in a completely different light: The reason that you must put in ten thousand or more hours of practice to become one of the world's best violinists or chess players or golfers is that the people you are being compared to or competing with have themselves put in ten thousand or more hours of practice. There is no point at which performance maxes out and additional practice does not lead to further improvement. So, yes, if you wish to become one of the best in the world in one of these highly competitive fields, you will need to put in thousands and thousands of hours of hard, focused work just to have a chance of equaling all of those others who have chosen to put in the same sort of work.

One way to think about this is simply as a reflection of the fact that, to date, we have found no limitations to the improvements that can be made with particular types of practice. As training techniques are improved and new heights of achievement are discovered, people in every area of human endeavor are constantly finding ways to get better, to raise the bar on what was thought to be possible, and there is no sign that this will stop. The horizons of human potential are expanding with each new generation.

5

Principles of Deliberate Practice on the Job

IT WAS 1968, AND THE VIETNAM WAR was in full swing. U.S. fighter pilots from the navy and air force were regularly engaging Soviet-trained North Vietnamese airmen flying Russian-made MiG fighter planes in dogfights, and the Americans weren't doing so well. In the previous three years, the pilots of both the navy and the air force had been winning about two-thirds of their dogfights: they downed two North Vietnamese jets for every one jet they lost. But in the first five months of 1968 the ratio for the navy pilots had dropped down to about one-to-one: the U.S. Navy had shot down nine MiGs, but lost ten of its own jets. Furthermore, over the summer of 1968, navy pilots had fired more than fifty air-to-air missiles without shooting down a single MiG. The navy's brass decided that something had to be done.

That something turned out to be the establishment of the now-famous Top Gun school, properly known as the U.S. Navy Strike Fighter Tactics Instructor Program (and originally the U.S. Navy Fighter Weapons School). The school would teach navy pilots how to fight more effectively and, it was hoped, increase their success rate in dogfights.

The program that the navy designed had many of the elements of deliberate practice. In particular, it gave the student pilots a chance to try different things in different situations, get feedback on their performance, and then apply what they had learned.

The navy picked its best pilots to be the trainers. These men would play the role of the enemy North Vietnamese pilots and engage the students in air-to-air "combat." The trainers, who were known collectively as the Red Force, flew fighter planes that were similar to the MiGs, and they used the same Soviet tactics the North Vietnamese pilots had learned. Thus they were, for all practical purposes, topnotch North Vietnamese fighter pilots, with one exception: instead of missiles and bullets, their aircraft were equipped with cameras to record each encounter. The dogfights were also tracked and recorded by radar.

The students who attended the Top Gun academy were the next best fighter pilots in the navy after the trainers, and collectively they were known as the Blue Force. They flew U.S. Navy fighter jets, again without the missiles or bullets. Each day they would climb into their planes and take off to face the Red Force. In those combats the pilots were expected to push their planes — and themselves — right up to the edge of failure in order to learn what the planes were capable of and what was required to get that performance out of them. They tried different tactics in different situations, learning how best to respond to what the other guys were doing.

The pilots of the Red Force, being the best the navy had, generally won the dogfights. And the trainers' superiority only increased over time, because every few weeks a whole new class of students would enter the Top Gun academy, while the trainers stayed there month after month, accumulating more and more dogfight experience as time went on and getting to the point at which they had seen pretty much everything the students might throw at them. For each new class the first few days of dogfights, in particular, were usually brutal defeats for the Blue Force.

That was okay, however, because the real action occurred once the pilots landed, in what the navy called "after-action reports." During these sessions the trainers would grill the students relentlessly: What did you notice when you were up there? What actions did you take? Why did you choose to do that? What were your mistakes? What could you have done differently? When necessary, the trainers could pull out the films of the encounters and the data recorded from the radar units and point out exactly what had happened in a dogfight. And both during and after the grilling the instructors would offer suggestions to the students on what they could do differently, what to look for, and what to be thinking about in different situations. Then the next day the trainers and students would take to the skies and do it all over again.

Over time the students learned to ask themselves the questions, as it was more comfortable than hearing them from the instructors, and each day they would take the previous session's lessons with them as they flew. Slowly they internalized what they'd been taught so that they didn't have to think so much before reacting, and slowly they would see improvement in their dogfights against the Red Force. And when the class was over, the Blue Force pilots — now much more experienced in dogfighting than almost any pilot who hadn't been to Top Gun — returned to their units, where they would become squadron training officers and pass on what they had learned to the other pilots in their squadrons.

The results of this training were dramatic. U.S. forces had stopped their bombing throughout all of 1969, so there were no dogfights that year, but the air war resumed in 1970, including air-to-air combat between fighters. Over the next three years, from 1970 to 1973, U.S. Navy pilots shot down an average of 12.5 North Vietnamese fighter planes for every U.S. Navy plane that was lost. During the same time, air force pilots had approximately the same two-to-one ratio they had had before the bombing halt. Perhaps the clearest way to see the results of the Top Gun training is to look at the "kills per engagement" statistics. Throughout the entire war, U.S. fighters downed an enemy jet an average of once every five encounters. However, in 1972, which was the last full year of fighting, Navy fighter pilots shot down an average of 1.04 jets per encounter. In other words, on average, every time navy pilots came in contact with the enemy they would down an enemy plane.

Noticing the dramatic effects of Top Gun training, the air force would later institute training exercises designed to prepare its own pilots for air-to-air combat, and both services continued this training after the end of the Vietnam War. By the time of the First Gulf War, both services had honed their programs so much that the pilots were far better trained than those in almost any other fighting service in the world. During the seven months of the First Gulf War, U.S. pilots shot down thirty-three enemy planes in air-to-air combat, losing only one plane in the process — perhaps the most dominant performance in combat aviation history.

The question that the navy had faced in 1968 is familiar to people in organizations and professions of almost any type: What is the best way to improve performance among people who are already trained and on the job?

In the navy's case, the problem was that the pilots' training hadn't truly prepared them to face other pilots in other jet fighters who were trying to shoot them down. Experience in other wars had shown that pilots who had won their first dogfight were much more likely to survive their second, and that the more dogfights a pilot fought and survived, the more likely he was to win the next one. Indeed, once a pilot had won twenty dogfights or so, he had almost a 100 percent chance of winning the next one and the one after that. The catch was, of course, that the cost of that sort of on-the-job training was unacceptably high. The navy was losing one plane for every two planes it managed to shoot down, and at one point it became an even trade — losing a plane for every enemy plane that was shot down. And with every plane that went down there was a pilot and, in the case of two-seater jets, a radio-intercept officer who might be killed or captured.

While there aren't too many fields in which the price of poor performance can be death or a prison camp, there are many in which the costs of mistakes can be unacceptably high. In medicine, for example, while doctors' lives aren't at stake, patients' lives can be. And in business situations a mistake can cost time, money, and future opportunities.

To its credit, the navy was able to devise a successful way to train its pilots without putting them in much danger. (Though not completely out of danger, of course. The training was so intense and close to the edge of the pilots' flying abilities that planes sometimes did crash and on rare occasions pilots did die, but it was far less likely than if the pilots had had to rely on on-the-job training.) Top Gun provided pilots with the opportunity to try different things and make mistakes without fatal consequences, to get feedback and figure out how to do better, and then to put their lessons to the test the next day. Over and over again.

It is never easy to design an effective training program, whether for fighter pilots or surgeons or business managers. The navy did it mainly through trial and error, as you find when you read histories of the Top Gun program. There was a debate, for instance, over how realistic the combat had to be, with some wanting to dial it back and lessen the risk to the pilots and the planes, and others arguing that it was important to push the pilots as hard as they would be pushed in real combat. Fortunately, the latter viewpoint eventually prevailed. We know now from studies of deliberate practice that the pilots learned best when they were pushed out of their comfort zones.

It has been my experience that there are many, many areas in the working world today where the lessons learned from studies of expert performers can help improve performance — in essence, to design Top Gun programs for different fields. I don't mean that literally, of course. No fighter jets, no six-g turns, no fancy nicknames like Maverick or Vi-per or Ice Man (unless you really want to). What I do mean is that if you follow the principles of deliberate practice you can develop ways to identify the top performers in a field and train other, lesser performers and bring them up closer to that top level. And by doing that it is possible to raise the performance level of an entire organization or profession.

PRACTICING WHILE GETTING WORK DONE

In the professional world, and especially in the corporate world, there's no shortage of people who make a living by offering advice on how to improve. They call themselves consultants or counselors or coaches, and they write books, give speeches, and lead seminars. They feed a seemingly insatiable appetite among their customers for anything that might provide a competitive edge. Of all the myriad approaches out there, the ones most likely to succeed are the ones that most resemble deliberate practice.

For several years I have been communicating with one of these individuals who has worked to understand the principles of deliberate practice and incorporate them into his training and coaching of corporate leaders. When Art Turock from Kirkland, Washington, first contacted me in 2008, much of our discussion centered on sprinting rather than corporate leadership. Art competes in masters-level track and field competitions, and I had gotten interested in how sprinters practice, in part because the great sprinter Walter Dix had been running for Florida State University, where I am based, so we had something in common from the beginning. Art had come across my name and a description of deliberate practice in an article in *Fortune* magazine, and as we talked I could tell that he was fascinated by the idea that deliberate practice could be applied equally well to business and running.

Since that first contact, Art has fully embraced the deliberate-practice mindset. He speaks of getting people out of their comfort zones in order to practice new skills and expand their abilities. He emphasizes the importance of feedback. And he studies the characteristics of some of the world's best business leaders—such as Jack Welch, the long-time chairman and chief executive officer of General Electric—to figure out what sorts of leadership, sales, and self-management skills other businesspeople should be developing to become elite performers.

His message to clients starts with mindset. The first step toward enhancing performance in an organization is realizing that improvement is possible only if participants abandon business-as-usual practices. Doing so requires recognizing and rejecting three prevailing myths.

The first is our old friend, the belief that one's abilities are limited by one's genetically prescribed characteristics. That belief manifests itself in all sorts of "I can't" or "I'm not" statements: "I'm just not very creative." "I can't manage people." "I'm not any good with numbers." "I can't do much better than this." But, as we've seen, the right sort of practice can help pretty much anyone improve in just about any area they choose to focus on. We can shape our own potential.

Art has a clever technique for making this point with his clients. When he is speaking with corporate leaders and he hears someone express one of these "I can't" or "I'm not" attitudes, he throws a red challenge flag like an NFL coach protesting a referee's call. It is meant to send the signal that the person who expressed the negative thoughts needs to reevaluate and revise them. The sudden appearance of a red flag in a conference room lightens the atmosphere, but it also makes his point in a way that people remember: mindset matters.

The second myth holds that if you do something for long enough, you're bound to get better at it. Again, we know better. Doing the same thing over and over again in exactly the same way is not a recipe for improvement; it is a recipe for stagnation and gradual decline.

The third myth states that all it takes to improve is effort. If you just try hard enough, you'll get better. If you want to be a better manager, try harder. If you want to generate more sales, try harder. If you want to improve your teamwork, try harder. The reality is, however, that all of these things — managing, selling, teamwork — are specialized skills, and unless you are using practice techniques specifically designed to improve those particular skills, trying hard will not get you very far.

The deliberate-practice mindset offers a very different view: anyone can improve, but it requires the right approach. If you are not improving, it's not because you lack innate talent; it's because you're not practicing the right way. Once you understand this, improvement becomes a matter of figuring out what the "right way" is.

And this, of course, is what Art Turock — like his many peers in the training and development world — has set out to do, but in Art's case much of the advice he offers has its roots in the principles of deliberate practice. One particular such approach is what Art calls "learning while real work gets done."

The approach acknowledges that businesspeople are so busy that they have hardly any time to practice their skills. They are in a totally different situation than, say, a concert pianist or pro athlete who spends relatively little time performing and thus can devote hours to practice each day. So Art set out to come up with ways that normal business activities could be turned into opportunities for purposeful or deliberate practice.

For example, a typical company meeting might have one person in front of a room giving a PowerPoint presentation, while managers and coworkers sit in the dark and try to stay awake. That presentation serves a normal business function, but Art makes the argument that it can be redesigned to serve as a practice session for everyone in the room. It might go like this: The speaker chooses a particular skill to focus on during the presentation — telling engaging stories, for example, or speaking more extemporaneously and relying less on the PowerPoint slides — and then tries to make that particular improvement during the presentation. Meanwhile, the audience takes notes on how the presenter's performance went, and afterward they practice giving feedback. If done just once, the presenter may get some useful advice, but it's not clear how much difference it will make, as any improvement from such a one-off session is likely to be minor. However, if the company makes it a regular practice in all staff meetings, employees can steadily improve on various skills.

Art has helped install this process at a number of companies, ranging from Fortune 500 corporations to midsize regional companies. One in particular, the Blue Bunny ice cream company, adopted the approach and even added its own twist. Its regional sales managers regularly visit the company's primary accounts — the grocery store chains and other businesses that sell lots of ice cream products — and several times a year each regional sales manager will meet with the company's senior sales managers to talk about strategy for an upcoming sales call. Traditionally these account reviews were just sales updates, but the company found a way to add a practice component. For the most challenging aspect of the upcoming sales call, the meeting is carried out as role-play, with the regional sales manager making his presentation to a colleague who pretends to be the account's primary buyer. After the presentation, the regional sales manager gets feedback from the other managers in the room, telling him what he did well and what he needs to change or improve. The next day the manager makes his presentation once more, again with feedback. Both practice rounds are videotaped so that the managers can view and review their performance. By the time the manager gives the actual presentation to the client, it has been polished and improved beyond what would have otherwise been possible.

One benefit of "learning while real work gets done" is that it gets people into the habit of practicing and thinking about practicing. Once they understand the importance of regular practice — and realize just how much they can improve by using it — they look for opportunities throughout the day in which normal business activities can be transformed into practice activities. Eventually, practicing becomes just a normal part of the business day. If it works as intended, the result is a completely different mindset from the usual one in which the business day is for work and practice is done only on special occasions, as when a consultant arrives and runs a training session. This practice-driven mindset is very similar to that of expert performers, who are constantly practicing and otherwise seeking ways to hone their skills.

For anyone in the business or professional world looking for an effective approach to improvement, my basic advice is to look for one that follows the principles of deliberate practice: Does it push people to get outside their comfort zones and attempt to do things that are not easy for them? Does it offer immediate feedback on the performance and on what can be done to improve it? Have those who developed the approach identified the best performers in that particular area and determined what sets them apart from everyone else? Is the practice designed to develop the particular skills that experts in the field possess? A yes answer to all those questions may not guarantee that an approach will be effective, but it will certainly make that much more likely.

THE TOP GUN APPROACH TO LEARNING

One of the major challenges facing anyone trying to apply the principles of deliberate practice is figuring out exactly what the best performers do that sets them apart. What are, in the words of one very successful book, the habits of highly effective people? In the business world and elsewhere, this is a difficult question to answer with any certainty.

Fortunately there's a way around this that can be used in a variety of situations. Think of it as the Top Gun approach to improvement. In the early days of the Top Gun project, no one stopped to try to figure out what made the best pilots so good. They just set up a program that mimicked the situations pilots would face in real dogfights and that allowed the pilots to practice their skills over and over again with plenty of feedback and without the usual costs of failure. That is a pretty good recipe for training programs in many different disciplines.

Consider the task of interpreting X-rays to detect breast cancer. When a woman gets her annual mammogram, the images are sent to a radiologist, who must examine them and determine if there are any abnormal areas of the breast that need to be tested further. In most cases the women coming in for the mammograms have no symptoms indicating the presence of breast cancer, so the X-ray images are all the radiologist has to go on. And research has found that, just as with the navy pilots during the early stages of the Vietnam War, some radiologists are able to perform this job much better than others. Tests have shown, for example, that some radiologists are much more accurate than others in distinguishing between benign and malignant lesions.

The main problem that radiologists face in this situation is the difficulty in getting effective feedback on their diagnoses, which limits how much they can improve over time. Part of the challenge is that only four to eight cases of cancer are expected to be found among every thousand mammograms. And even when the radiologist detects what may be cancer, the results are sent back to the patient's personal doctor, and the radiologist will seldom be notified about the biopsy's results. It is even less common for the radiologist to be informed whether the patient develops breast cancer within the next year or so after the mammogram — which would give the radiologist a chance to reexamine the mammograms and see whether he or she might have missed the early signs of cancer.

With little chance for the type of feedback-driven practice that leads to improvement, radiologists do not necessarily get better with more experience. A 2004 analysis of half a million mammograms and 124 American radiologists was not able to identify any background factors of the radiologists, such as years of experience or the number of yearly diagnosed mammograms, that were related to accuracy of diagnosis. The authors of that study speculated that the differences in performance among the 124 radiologists might be due to the initial training the doctors received before starting independent practice.

After completing medical school and their internships, future radiologists have a four-year specialty training program in which they learn their craft by working with experienced radiologists who teach them what to look for and allow them to read mammograms of patients. These supervisors check their readings, telling them whether their diagnoses and identification of abnormal areas agreed with the supervisor's own expert opinion. Of course, there is no way to know immediately whether the supervisor was right or wrong, and even experienced radiologists are estimated to miss one cancer per every thousand readings and to regularly request unnecessary biopsies.

In the published version of my keynote at the 2003 annual meeting of the American Association of Medical Colleges, I suggested a Top Gun–like approach to training radiologists to interpret mammograms more effectively. The main problem, as I saw it, was that radiologists do not have the chance to practice their readings over and over again, getting accurate feedback with each attempt. So this is what I suggested: You'd start by collecting a library of digitized mammograms taken from patients years ago along with enough information from those patients' records to know the ultimate outcome — whether there actually was a cancerous lesion present and, if so, how the cancer progressed over time. In this way we would collect, in essence, a number of test questions in which the answer is known: Is cancer present or not? Some of the images would be from women who never did develop cancer, while others would be from women whose doctors correctly diagnosed cancer from the image. It would even be possible to include images in which cancer was present but the doctor missed it originally, as long as a retrospective analysis of the image discovered signs of the cancer's presence. Ideally, the images would be chosen for their training value. There would be little value, for instance, in having lots of images of clearly healthy breasts or breasts with obvious tumors; the best images would be those that would challenge the radiologists by displaying cancerous or benign abnormalities.

Once such a library was accumulated, it could easily be turned into a training tool. A simple computer program could be written to let radiologists work through the images, making diagnoses and getting feedback. The program could respond to a wrong answer by displaying other images with similar characteristics so that the doctor could get more practice on his or her weaknesses. This is no different in theory than a music teacher noticing that a student is having difficulty with a particular type of finger movement and assigning a series of exercises designed to improve that movement. It would be, in short, deliberate practice.

I am delighted to report that a digital library very similar to what I proposed has been built in Australia; it allows radiologists to test themselves with a variety of mammograms they can retrieve from the library. A 2015 study reported that performance on a test set of mammograms from the library predicted how accurately radiologists interpreted mammograms in their professional practice. The next step will be to demonstrate that improvements from training with the library lead to increased accuracy in the clinic.

A similar library has been independently assembled for pediatric ankle X-rays. As reported in a 2011 study, a group of doctors at Morgan Stanley Children's Hospital in New York City assembled a set of 234 cases of possible ankle injury in a child. Each case included a series of X-rays and a brief summary of the patient's history and symptoms. The doctors used this library to train radiology residents. A resident would be given the case details and the X-rays and asked to make a diagnosis — in particular, to classify the case as normal or abnormal and, if it was abnormal, to point out the abnormality. Immediately afterward the resident would get feedback on the diagnosis from an experienced radiologist, who would explain what was right and wrong about the diagnosis and what was missed.

The doctors running the study found that this practice and feedback helped the residents improve their diagnostic powers dramatically. At first the residents were relying on their own previous knowledge, and their diagnoses were hit or miss, but after twenty or so trials the effects of the regular feedback started to kick in, and the residents' accuracy began to increase steadily. The improvement continued through all 234 cases, and from all appearances would have continued through at least several hundred more if they'd been available.

In short, this sort of training with immediate feedback — either from a mentor or even a carefully designed computer program — can be an incredibly powerful way to improve performance. Furthermore, I believe that radiology training could be even more effective if an effort were made up front to determine what sorts of issues are likely to cause problems for new radiologists and design the training to focus more on those issues — in essence, to learn more about the role that mental representations play in making accurate diagnoses and apply that understanding in designing the training.

Some researchers have used the same sort of think-aloud protocol I used in studying Steve Faloon in order to understand the mental processes underlying superior performance among radiologists, and it seems clear from this work that the best radiologists have indeed developed more accurate mental representations. We even have a good idea of the types of cases and lesions that give trouble to the less expert radiologists. Unfortunately, we do not yet know enough about the differences between how expert and nonexpert radiologists make their judgments to design training programs to address the weaknesses of the less accomplished ones.

However, we can see exactly how this sort of training might work in the case of laparoscopic surgery, for which researchers have done much more work in assembling an understanding of the types of mental representations that effective doctors use in their work. In one study, a group led by Lawrence Way, a surgeon at the University of California, San Francisco, set out to understand what led to a particular type of injury to patients' bile ducts during laparoscopic surgery to remove their gallbladders. In almost every case, these injuries were due to what the group called a "visual perceptual illusion"—that is, the surgeon mistook one body part for another. This led the surgeon to cut a bile duct instead of, say, a cystic duct, which was the intended target. The misperception was so strong that even when a surgeon noticed an anomaly, he would often keep going without stopping to question whether something might be wrong. Other researchers studying factors in the success of laparoscopic surgery have found that expert surgeons develop ways of getting a clearer look at the body parts, such as pushing some tissues aside in order to produce a better view for the camera that is used to guide the procedure.

This is exactly the sort of information that makes it possible to improve performance through designed practice. Knowing what the best laparoscopic surgeons do right, and knowing the most common mistakes, it should be possible to design training exercises outside the operating room to improve surgeons' mental representations. One way would be to use videos of actual surgeries, run them up to a certain decision point, then stop them and ask, "What do you do next?" or "What are you looking at here?" The answer might be a line on the video screen showing where to cut, or drawing the outline of a bile duct, or a suggestion to push aside a bit of tissue to get a better look. The surgeons could get immediate feedback on their answers, going back to correct erroneous thinking and moving on to other, perhaps tougher challenges when they do it right.

Using such an approach, doctors could carry out dozens or hundreds of training runs, focusing on various aspects of the surgery that are known to cause problems, until they develop effective mental representations.

More generally, this Top Gun approach could be applied in a wide variety of areas in which people could benefit from practicing something over and over again "offline"—that is, away from their real job, where mistakes have real consequences. This is the rationale behind using simulators to train pilots, surgeons, and many others in high-stakes professions. Indeed, using libraries of mammograms to train radiologists is a kind of simulation. But there are many more areas where this concept could be put to use. One could imagine, for instance, creating a library of case studies designed to help tax accountants hone their skills in certain specialties, or to help intelligence analysts improve their abilities to interpret what is happening in a foreign country.

Even in those areas where simulators or other techniques are already being used to improve performance, their effectiveness could be greatly increased by explicitly taking into account the lessons of deliberate practice. As I mentioned, while simulators are used in a number of areas of surgery, they could probably improve performance much more effectively if their design took into account what is known — or what can be learned — about the mental representations of the most effective surgeons in a given specialty. It is also possible to improve simulator training by determining which errors are most common and most dangerous and by setting up the simulators to focus on the situations where those errors happen. For example, during surgery it is not uncommon that some interruption brings the procedure to a temporary halt, and if the interruption occurs while someone is starting to check the blood type prior to a blood transfusion, it is critical that the person continue this checking when the activity resumes after the interruption. To help surgeons and other members of the medical team gain experience in dealing with such interruptions, a simulator supervisor can initiate an interruption at exactly the critical point on various occasions. The possibilities for such sorts of simulator practice are endless.

KNOWLEDGE VERSUS SKILLS

One of the implicit themes of the Top Gun approach to training, whether it is for shooting down enemy planes or interpreting mammograms, is the emphasis on *doing.* The bottom line is what you are able to do, not what you know, although it is understood that you need to know certain things in order to be able to do your job.

This distinction between knowledge and skills lies at the heart of the difference between traditional paths toward expertise and the deliberate-practice approach. Traditionally, the focus is nearly always on knowledge. Even when the ultimate outcome is being able to do something — solve a particular type of math problem, say, or write a good essay — the traditional approach has been to provide information about the right way to proceed and then mostly rely on the student to apply that knowledge. Deliberate practice, by contrast, focuses solely on performance and how to improve it.

When Dario Donatelli, the third person to take part in my memory experiment at Carnegie Mellon, began trying to improve his digit memory, he spoke with Steve Faloon, who told him exactly what he had done to get up to eighty-two digits. Indeed, Dario and Steve were friends and saw each other regularly, so Steve often gave Dario ideas and instruction about how to create mnemonics for groups of digits and how to organize those groups in his memory. In short, Dario had a tremendous amount of *knowledge* about how to memorize digits, but he still had to develop the *skill.* Because Dario didn't have to rely on the same trial and error that Steve did, he was able to improve more quickly, at least in the beginning, but it was still a long, slow process to develop his memory. The knowledge helped, but only to the extent that Dario had a better idea of how to practice in order to develop the skill.

When you look at how people are trained in the professional and business worlds, you find a tendency to focus on knowledge at the expense of skills. The main reasons are tradition and convenience: it is much easier to present knowledge to a large group of people than it is to set up conditions under which individuals can develop skills through practice.

Consider medical training. By the time future doctors have graduated from college, they've spent more than a decade and a half being educated, but almost all of that education will have focused on providing knowledge, little or none of which will have direct application to the skills they will need as doctors. Indeed, future doctors don't begin their training in medicine until they enter medical school, and even once they reach that milestone they spend a couple of years in coursework before getting to clinical work, where they finally start developing their medical skills. It is mainly once they graduate from medical school that they begin specializing and developing the necessary skills for surgery or pediatrics or radiology or gastroenterology or whatever specialty they choose. And only at this point, when they're interns and residents working under the supervision of experienced doctors, do they finally learn many of the diagnostic and technical skills that they need for their specialties.

After their internships and residencies, some doctors get a fellowship to continue on with even more specialized training, but that is the end of their official supervised training. Once new doctors have reached this stage, they go to work as full-fledged physicians with the assumption that they've developed all the skills they need to treat patients effectively.

If this all sounds vaguely familiar, it should, for it is very similar to the pattern I described in chapter 1 when explaining how one might learn to play tennis: take some tennis lessons, develop enough skill to play the game competently, and then set aside the intense training that characterized the original learning period. As I noted, most people assume that as you continue to play tennis and accumulate all of those hours of "practice," you will inevitably get better, but the reality is different: as we've seen, people generally don't get much better just by playing the game itself, and, sometimes, they'll actually be worse.

This similarity between doctors and recreational tennis players was shown in 2005 when a group of researchers at Harvard Medical School published an extensive review of research looking at how the quality of care that doctors provide changes over time. If years of practice make physicians better, then the quality of care they give should increase as they amass more experience. But just the opposite was true. In almost every one of the five dozen studies included in the review, doctors' performance grew worse over time or, at best, stayed about the same. The older doctors knew less and did worse in terms of providing appropriate care than doctors with far fewer years of experience, and researchers concluded that it was likely the older doctors' patients fared worse because of it. Only two of sixty-two studies had found doctors to have gotten better with experience. Another study of decision-making accuracy in more than ten thousand clinicians found that additional professional experience had only a very small benefit.

Not surprisingly, the same thing is true for nurses as well. Careful studies have shown that very experienced nurses do not, on average, provide any better care than nurses who are only a few years out of nursing school.

We can only speculate as to why the performance of older and more experienced health-care providers is not consistently better — and is sometimes worse — than that of their younger, less experienced peers. Certainly the younger doctors and nurses will have received more up-to-date knowledge and training in school, and if continuing education doesn't keep doctors effectively updated, then the older they get, the less current their skill will be. But one thing is clear: with few exceptions, neither doctors nor nurses gain expertise from experience alone.

Of course, doctors do work very hard to improve. They're constantly attending conferences, meetings, workshops, minicourses, and the like, where the goal is updating them on the latest thinking and techniques in their fields. As I was writing this, I visited the website doctorsreview.com, which bills itself as "the most complete medical meetings listings on the web." On the meetings search page I picked a field at random — cardiology — and a month at random — August 2015 — and then hit a button to request a list of all the meetings on that topic during that month. It gave me twenty-one, ranging from the Cardiovascular Fellows' Bootcamp in Houston to Ultrasound-

Guided Vascular Access in St. Petersburg, Florida, and Electrophysiology: Arrhythmias Unraveled for Primary Care Providers and Cardiologists in Sacramento, California. And that was just one month for one specialty. The site promised more than twenty-five hundred meetings in all.

In short, doctors are clearly serious about keeping their skills sharp. Unfortunately, the way they have been doing it just isn't working. Several researchers have examined the benefits of continuing medical education for practicing physicians, and the consensus is that while it is not exactly worthless, it's not doing much good, either. But to the credit of the medical profession, I have found doctors to be exceptionally willing to look for shortcomings in their fields and search for ways to correct them. It is largely because of this willingness that I have spent so much time working with doctors and other medical professionals. It is not because medical training is less effective than training in other fields, but rather because those in this field are so motivated to find ways to improve.

Some of the most compelling research on the effectiveness of continuing professional education for physicians has been done by Dave Davis, a doctor and educational scientist at the University of Toronto. In a very influential study, Davis and a group of colleagues examined a wide-ranging group of educational "interventions," by which they meant courses, conferences and other meetings, lectures, and symposia, taking part in medical rounds, and pretty much anything else whose goal was to increase doctors' knowledge and improve their performance. The most effective interventions, Davis found, were those that had some interactive component — role-play, discussion groups, case solving, hands-on training, and the like. Such activities actually did improve both the doctors' performance and their patients' outcomes, although the overall improvement was small. By contrast, the least effective activities were "didactic" interventions — that is, those educational activities that essentially consisted of doctors listening to a lecture — which, sadly enough, are by far the most common types of activities in continuing medical education. Davis concluded that this sort of passive listening to lectures had no significant effect at all on either doctors' performance or on how well their patients fared.

That study reviewed other studies of continuing medical education that had been published before 1999. A decade later, a group of researchers led by Norwegian researcher Louise Forsetlund updated Davis's work, looking at forty-nine new studies of continuing medical education that had been published in the meantime. That group's conclusions were similar to Davis's: continuing medical education can improve doctors' performance, but the effect is small, and the effects on patient outcomes are even smaller. In addition, it is mainly those education approaches with some interactive component that have an effect; lectures, seminars, and the like do little or nothing to help doctors improve their practice. Finally, the researchers found that no type of continuing medical education is effective at improving complex behaviors, that is, behaviors that involve a number of steps or require considering a number of different factors. In other words, to the extent that continuing medical education is effective, it is effective in changing only the most basic things that doctors do in their practices.

From the perspective of deliberate practice, the problem is obvious: attending lectures, minicourses, and the like offers little or no feedback and little or no chance to try something new, make mistakes, correct the mistakes, and gradually develop a new skill. It's as if amateur tennis players tried to improve by reading articles in tennis magazines and watching the occasional YouTube video; they may believe they're learning something, but it's not going to help their tennis game much. Furthermore, in the online interactive approaches to continuing medical education, it is very difficult to mimic the sorts of complex situations that doctors and nurses encounter in their everyday clinical practice.

Once they've finished their training, medical and other professionals are supposed to be able to work independently; they have no one assigned to them to play the role of the tennis pro, working with them to identify their weaknesses, coming up with training regimens to fix them, and then overseeing and even leading that training. More generally, the field of medicine — as is the case with most other professional fields — lacks a strong tradition of supporting the training and further improvement of practicing professionals. It is assumed that medical professionals are able to figure out, on their own, effective practice techniques and apply them to improve their performance. In short, the implicit assumption in medical training has been that if you provide doctors with the necessary knowledge — in medical school, through medical journals, or through seminars and continuing medical education classes — this should be sufficient.

There is a saying in medicine about learning surgical procedures that can be traced back to William Halsted, a surgical pioneer in the early twentieth century: "See one, do one, teach one." The idea is that all surgical trainees need in order to be able to carry out a new surgery is to see it done once and, after that, they can figure out how to do it by themselves on succeeding patients. It is the ultimate profession of faith in knowledge versus skill.

However, this faith was severely challenged in the 1980s and 1990s with the spread of laparoscopic surgery, or keyhole surgery, in which a surgery is performed with instruments inserted through a small opening in the body that may be well away from the surgical site. It required techniques that were radically different from those of traditional surgery, yet the general assumption was that experienced surgeons should be able to pick up this new technique relatively quickly without extra training. After all, they had all the necessary knowledge to carry out the procedures. However, when medical researchers compared the learning curves of surgeons who had a great deal of experience in traditional surgery with the learning curves of surgical trainees, they found no difference in how quickly the two groups mastered laparoscopic surgery and reduced the numbers of complications.

In short, neither their greater knowledge nor their greater experience in traditional surgery gave the experienced surgeons an advantage in developing skill in laparoscopic surgery. That skill, it turns out, must be developed independently. Because of these findings, surgeons today who wish to perform laparoscopic procedures must go through training supervised by expert laparoscopic surgeons and be tested on this specific skill.

It is not just the medical profession that has traditionally emphasized knowledge over skills in its education. The situation is similar in many other professional schools, such as law schools and business schools. In general, professional schools focus on knowledge rather than skills because it is much easier to teach knowledge and then create tests for it. The general argument has been that the skills can be mastered relatively easily if the knowledge is there. One result is that when college students enter the work world, they often find that they need a lot of time to develop the skills they need to do their job. Another result is that many professions do no better a job than medicine — and in most cases, a worse job — of helping practitioners sharpen their skills. Again, the assumption is that simply accumulating more experience will lead to better performance.

As is the case in so many situations, once you have figured out the right question to ask, you are halfway to the right answer. And when referring to improving performance in a professional or business setting, the right question is, How do we improve the relevant skills? rather than, How do we teach the relevant knowledge?

A NEW APPROACH TO TRAINING

As we've seen with the Top Gun approach and Art Turock's work, there are ways to immediately apply the principles of deliberate practice to improve skills in professional and business settings. But over the long term I believe the best approach will be to develop new skills-based training programs that will supplement or completely replace the knowledge-based approaches that are the norm now in many places. This strategy acknowledges that because what is ultimately most important is what people are able to do, training should focus on doing rather than on knowing—and, in particular, on bringing everyone's skills closer to the level of the best performers in a given area.

Since 2003 I've been working with medical professionals to show how deliberate practice could sharpen the skills that doctors rely on every day. Switching over to such methods would represent a paradigm shift and would have far-reaching benefits to doctors' abilities and, ultimately, to the health of their patients. In one very relevant study John Birkmeyer and colleagues invited a group of bariatric surgeons in Michigan to submit videotapes of typical examples of laparoscopic gastric bypasses from their clinics. The researchers then had the tapes evaluated anonymously by experts to assess the technical skills of the surgeons. For our purposes, the key finding was that there were large differences in patient outcomes among surgeons of differently rated technical skills, with patients of the more technically adept surgeons being less likely to experience complications or mortality. This suggests that patients could benefit greatly if less technically skilled surgeons could be helped to improve their skills. The results led to the creation of a project in which highly skilled surgeons coach less technically skilled surgeons to help them improve.

In the rest of this chapter I'll sketch out how the principles of deliberate practice could be applied to develop new and more effective training methods for doctors that should ultimately result in better outcomes for patients.

The first step is to determine with some certainty who the expert doctors are in a given area. How can we identify those doctors whose performance is reliably superior to other doctors? This is not always easy, as we discussed in chapter 4, but there are generally ways to do it with reasonable objectivity.

Because the bottom line in medicine is the health of the patient, what we really want to find is some patient outcome that can be definitively linked to the behavior of a doctor. This can be tricky because medical care is a complicated process involving many steps and many people, and there are relatively few outcome measures that can be clearly linked to the contributions of an individual caregiver. Nonetheless, there are at least two good examples that illustrate in general how we might identify expert doctors.

In 2007, a group of researchers led by Andrew Vickers of the Memorial Sloan Kettering Cancer Center in New York City reported the outcomes for nearly eight thousand men with prostate cancer who had had their prostates surgically removed. The procedures had been done by seventy-two different surgeons at four medical centers between 1987 and 2003. The goal of such surgeries is to remove the entire prostate gland along with any cancer in the surrounding tissue. This complex operation requires painstaking care and skill, and if it is not done exactly right, cancer is more likely to recur. Thus, the success rate in preventing a recurrence of cancer after this operation should offer an objective measure that can separate the best surgeons from the rest.

And this is what Vickers and his colleagues found: There was a major difference in skill between surgeons who had had a great deal of experience with this surgery and those who had relatively few such operations under their belts. While surgeons who had performed just 10 prostatectomies had a five-year cancer recurrence rate of 17.9 percent, those who had performed 250 prior surgeries had a recurrence rate of just 10.7 percent. In other words, you were almost twice as likely to have your cancer come back within five years if you were operated on by an inexperienced surgeon than if you were operated on by an experienced one. In a follow-up study, Vickers examined what happened to recurrence rates as surgeons got even more experience, and he found that the rates continued to drop up until the point where a surgeon had carried out 1,500 to 2,000 surgeries. At that point the surgeons had become essentially perfect at preventing five-year recurrence in the simpler cases where the cancer had not spread outside the prostate, while they were preventing recurrence in 70 percent of the more complex cases where the cancer had spread outside the prostate. After that, the success rate did not improve with more practice.

In the paper describing the results, Vickers noted that his group had not had the chance to figure out just what the highly experienced surgeons were doing differently. It seemed clear, however, that performing hundreds or thousands of the surgeries had led the doctors to develop particular skills that made a huge difference to their patients' outcomes. It is also worth noting that because increasing experience in the surgery led to increasing competence, there must have been some sort of feedback available to the surgeons that allowed them to improve over time by correcting and sharpening their techniques.

Surgery is different from most other areas of medicine in that many problems are immediately apparent, such as a rupture of a blood vessel or damage to tissue, and thus surgeons get immediate feedback about at least some of their mistakes. In the postoperative surgery suite, the patient's condition is monitored carefully. Occasionally at this stage there is bleeding or some other problem, and the patient must undergo surgery to correct the problem. Such corrective surgeries also give surgeons feedback about potentially avoidable problems. In the case of surgeries to remove cancerous lesions, laboratory analysis of the removed cancer tissue permits an analysis of whether all of the cancer was successfully removed. Ideally, all of the removed tissue should have some healthy tissue surrounding the cancer, and if the surgeon failed to provide these "clean margins," this provides yet another type of feedback he or she can use when carrying out similar operations in the future. In heart surgery, it is possible to test the repaired heart to assess the success of the surgery and to determine, if the surgery was not successful, what went wrong. Feedback like this is most likely the reason that surgeons, unlike most other medical professionals, get better as they gain experience.

Deliberate practice–based techniques to build surgical skills could be particularly valuable because it is clear from this study and others like it that surgeons need years and many surgeries to reach a point at which they can be considered experts. If training programs could be developed that halved the time a surgeon needs to reach expert status, it could make a major difference to patients.

A pattern of improvement similar to what Vickers observed in surgeons was seen in a study of radiologists interpreting mammograms. The radiologists improved considerably in their interpretations over the first three years they spent on the job, coming up with fewer and fewer false positives — that is, cases in which women did not have breast cancer but were called in for further screening — and then their rate of improvement slowed down sharply. Interestingly, this improvement over the first three years occurred only for radiologists who had not had fellowship training in radiology. Those doctors who had gone through a radiology fellowship did not have the same sort of learning curve but instead took only a few months on the job to reach the same skill level that the nonfellowship radiologists took three years to develop.

If the training received in fellowships helps radiologists reach expert status that much more quickly than they normally would, it seems reasonable to assume that a well-designed training program that didn't require a fellowship might be able to accomplish the same thing.

Once you have identified people who consistently perform better than their peers, the next step is to figure out what underlies that superior performance. This usually involves some variation of the approach, described in chapter 1, that I used in the memory work with Steve Faloon. That is, you get retrospective reports, you have people describe what they're thinking about as they perform a task, and you observe which tasks are easier or harder for someone and draw conclusions from that. The researchers who have studied doctors' thought processes in order to understand what separates the best from the rest have used all of these techniques.

A good example of this approach is a recent study of eight surgeons who were quizzed about their thought processes before, during, and after they performed laparoscopic surgeries. These surgeries, which are carried out with a small incision through which the surgical instruments are inserted and then guided to the surgical target, require a great amount of preparation and the ability to adapt to whatever conditions are discovered once the surgery begins. A major aim of the study was to identify the sorts of decisions the surgeons made throughout the process and to figure out how they made those decisions. The researchers listed several types of decisions that the surgeons had to make during surgery, such as which tissues to cut, whether to switch from laparoscopy to open surgery, and if they needed to drop their original surgical plan and improvise.

The details are mostly interesting only to laparoscopic surgeons and those who teach them, but one finding has much broader relevance. Relatively few of the surgeries were simple and straightforward enough to be carried out by following the basic pattern one expects for such surgeries; to the contrary, most of them took some unexpected turn or threw up some unexpected obstacle that forced the surgeon to think carefully about what he or she was doing and make some sort of decision. As the researchers who performed the study put it, "even expert surgeons find themselves in situations in which they must thoughtfully reevaluate their approach during surgery, evaluating alternative actions, such as the selection of different instruments or changing the position of the patient."

This ability — to recognize unexpected situations, quickly consider various possible responses, and decide on the best one — is important not just in medicine but in many areas. For instance, the U.S. Army has spent a considerable amount of time and effort figuring out the best way to teach what it calls "adaptive thinking" to its officers, particularly the lieutenants, captains, majors, and colonels who are on the ground with the troops and may have to immediately determine the best actions in response to an unexpected attack or other unforeseen event. It has even developed the Think Like a Commander Training Program to teach this sort of adaptive thinking to its junior officers using deliberate-practice techniques.

Research into the mental processes of the best doctors has shown that while they may have prepared surgical plans before they start, they regularly monitor the surgeries in progress and are ready to switch gears if necessary. This was apparent from a series of recent studies done by medical researchers in Canada who observed operations that the surgeons had predicted would be challenging. When the researchers interviewed the surgeons after the operations about their thought processes during the surgery, they found that the main way the surgeons detected problems was by noticing that something about the surgery didn't match the way they had visualized the surgery in their preoperative plan. Once they noticed the mismatch, they came up with a list of alternative approaches and decided which was most likely to work.

This points to something important about how these experienced surgeons perform: over time they have developed effective mental representations that they use in planning the surgery, in performing it, and in monitoring its progress so that they can detect when something is wrong and adapt accordingly.

Ultimately, then, if we are going to understand what makes a superior surgeon, we need to have a good idea of what a superior surgeon's mental representations look like. Psychologists have developed various ways to study mental representations. One standard approach for examining the mental representations that people use to guide themselves through a task is to stop them in the middle of the task, turn out the lights, and then ask them to describe the current situation, what has happened, and what is about to happen. (We saw an example of this method in the research on soccer players described in chapter 3.) This obviously won't work for surgeons in an operating room, but there are other ways to investigate the mental representations of people in potentially risky situations like surgery. In cases where simulators are available — flight training, for instance, or certain types of medical procedures — it is actually possible to stop in the middle and quiz people. Or, in the case of real surgeries, the doctors can be questioned before and after the operations about how they envision the surgery going and about their thought processes during the surgery; in this case, it's best to combine the interviews with observations of the surgeons' actions during the surgery. Ideally, you would like to identify characteristics of mental representations that are associated with greater success in the surgeries.

It has been primarily since the beginning of the twenty-first century that a few researchers have been successful in identifying those practicing doctors with reliably superior performance and have begun to investigate their mental processes. It's already clear, however, that a major factor underlying the abilities of the world's best doctors is the quality of their mental representations. This implies that a major part of applying the lessons of deliberate practice to medicine will be finding ways to help doctors develop better mental representations through training — a situation that holds in most other professions as well.

6

Principles of Deliberate Practice in Everyday Life

IN 2010 I GOT AN E-MAIL from a man named Dan McLaughlin from Portland, Oregon. He had read about my deliberate-practice research in various places, including Geoff Colvin's book *Talent Is Overrated,* and he wanted to use it in his efforts to become a professional golfer.

To understand just how audacious this was, you need to know a little about Dan. He had not played on his high school or college golf team. In fact, he'd never really played golf at all. He'd been to a driving range with friends a few times, but he'd never played a full eighteen-hole round of golf in his life. Indeed, at thirty years old, he had never been a competitive athlete of any sort.

But he had a plan, and he was serious about it: he would quit his job as a commercial photographer and spend the next six or so years learning to play golf. Having read Malcolm Gladwell's book *Outliers* and taken "the ten-thousand-hour rule" at face value, Dan figured he would put in ten thousand hours of deliberate practice and become a good enough player to join the Professional Golfers' Association tour. To get on the tour, he'd have to first gain admission to the PGA Tour Qualifying Tournament and then do well enough in that tournament to receive a PGA Tour card. This would allow him to compete in PGA tournaments.

A year and a half after starting his project, which he called "the Dan Plan," he gave an interview to *Golf* magazine. When the writer asked him why he was doing it, Dan gave an answer I really liked. He said he didn't appreciate the attitude that only certain people can succeed in certain areas — that only those people who are logical and "good at math" can go into mathematics, that only athletic people can go into sports, that only musically gifted people can become really good at playing an instrument. This sort of thinking just gave people an excuse not to pursue things that they might otherwise really enjoy and perhaps even be good at, and he didn't want to fall into that trap. "That inspired me to try something completely different from anything I'd ever done," he said. "I wanted to prove that anything's possible if you're willing to put in the time."

Even more than this statement, I liked Dan's realization that deliberate practice isn't just for kids who are beginning a life of training to become chess grandmasters or Olympic athletes or world-class musicians. Nor is it just for members of large organizations, like the U.S. Navy, that can afford to develop some high-intensity training program. Deliberate practice is for everyone who dreams. It's for anyone who wants to learn how to draw, to write computer code, to juggle, to play the saxophone, to pen "the Great American Novel." It's for everyone who wants to improve their poker game, their softball skills, their salesmanship, their singing. It's for all those people who want to take control of their lives and create their own potential and not buy into the idea that this right here, right now, is as good as it gets.

This chapter is for them.

FIRST, FIND A GOOD TEACHER

Another of my favorite correspondents is Per Holmlöv, a Swedish man who started taking karate lessons when he was sixty-nine years old. He set himself a goal of gaining a black belt by the time he was eighty. Per wrote to me after he'd been training for about three years. He told me that he thought he was progressing too slowly, and he asked for some advice on how he could train more effectively.

Although he had been physically active all of his life, this was his first experience with martial arts. He was training in karate five or six hours a week and spending another ten hours a week in other exercise, mainly jogging in the woods and going to the gym. What more could he do?

When hearing about Per, some people's natural reaction might be, "Well, of course he's not progressing too quickly—he's seventy-two years old!" But that wasn't it. No, he wasn't going to improve as quickly as a twenty-four-year-old or even a fifty-four-year-old, but there was no doubt that he could get better more quickly than he was doing. So I offered some advice—the same advice I would offer to that twenty-four-year-old or that fifty-four-year-old.

Most karate training is done in a class with a number of students and a single instructor who demonstrates a movement, which the class imitates. Occasionally, the instructor might notice a particular student performing the movement incorrectly and offer a little one-on-one tutoring. But such feedback is rare.

Per was taking just such a class, so I suggested he get some personal sessions with a coach who could give advice tailored to Per's performance.

Given the expense of private instruction, people will often try to make do with group lessons or even YouTube videos or books, and those approaches will generally work to some degree. But no matter how many times you watch a demonstration in class or on YouTube, you are still going to miss or misunderstand some subtleties — and sometimes some things that are not so subtle — and you are not going to be able to figure out the best ways to fix all of your weaknesses, even if you do spot them.

More than anything else, this is a problem of mental representations. As we discussed in chapter 3, one of the main purposes of deliberate practice is to develop a set of effective mental representations that can guide your performance, whether you are practicing a karate move, playing a piano sonata, or performing surgery. When you're practicing by yourself, you have to rely upon your own mental representations to monitor your performance and determine what you might be doing wrong. This is not impossible, but it is much more difficult and less efficient than having an experienced teacher watching you and providing feedback. It is particularly difficult early in the learning process, when your mental representations are still tentative and inaccurate; once you have developed a foundation of solid representations, you work from those to build new and more effective representations on your own.

Even the most motivated and intelligent student will advance more quickly under the tutelage of someone who knows the best order in which to learn things, who understands and can demonstrate the proper way to perform various skills, who can provide useful feedback, and who can devise practice activities designed to overcome particular weaknesses. Thus, one of the most important things you can do for your success is to find a good teacher and work with him or her.

How do you find a good teacher? This process will likely entail some trial and error, but there are a few ways you can improve your chances of success. First, while a good teacher does not have to be one of the best in the world, he or she should be accomplished in the field. Generally speaking, teachers will only be able to guide you to the level that they or their previous students have attained. If you're a flat-out beginner, any reasonably skilled teacher will do, but once you've been training for a few years, you'll need a teacher who is more advanced.

A good teacher should also have some skill and experience in teaching in that field. Many accomplished performers are terrible teachers because they have no idea how to teach. Just because they themselves can do it doesn't mean they can teach others how to do it. Ask about a teacher's experience and, if possible, investigate and even talk to the teacher's former or current students. How good are they? How much of their skill can be attributed to that particular teacher? Do they speak highly of the teacher? The best students to talk to are those who started working with a teacher when they were at about the same level you are now, since their experience will be closest to what you yourself will get from a teacher. Ideally you want to find students similar in age and relevant experience. A teacher might be great with children and adolescents but have less experience with and understanding of how to help someone a few decades older.

When looking into a teacher's reputation, keep in mind the shortcomings of subjective judgments. Online rating sites are particularly vulnerable to these shortcomings, as the ratings on these sites often reflect how personable teachers are or how enjoyable it is to learn from them rather than how effective they are. In reading reviews of an instructor, skip over the stuff about how much fun their lessons are and look for specific descriptions of progress the students have made and obstacles they have overcome.

It's particularly important to query a prospective teacher about practice exercises. No matter how many sessions a week you have with an instructor, most of your effort will be spent practicing by yourself, doing exercises that your teacher has assigned. You want a teacher who will guide you as much as possible for these sessions, not only telling you what to practice on but what particular aspects you should be paying attention to, what errors you have been making, and how to recognize good performance. Remember: one of the most important things a teacher can do is to help you develop your own mental representations so that you can monitor and correct your own performance.

Dan McLaughlin of the Dan Plan offers a good — if extreme — example of how to use instructors to improve. Dan had read about deliberate practice and had absorbed many of its lessons, so from the beginning of his quest he understood the importance of personal instruction. Before he even began, he'd already recruited three instructors: a golf coach, a strength-and-conditioning coach, and a nutritionist.

Dan's later experience illustrates one final lesson about instruction: you may need to change teachers as you yourself change. For several years he improved with his original golf coach, but at some point he stopped getting better. He had absorbed everything this coach could teach him, and he was ready to find a coach at the next level. If you find yourself at a point where you are no longer improving quickly or at all, don't be afraid to look for a new instructor. The most important thing is to keep moving forward.

ENGAGEMENT

Returning to Per's story, we can see another essential element of deliberate practice that benefits from the right sort of one-on-one instruction: engagement. I suspected that his group karate lessons were failing to keep him fully focused and engaged. In group classes, with the instructor at the front and all of the students following en masse, it is far too easy to just "go through the motions" instead of actually practicing them with the specific goal of improving a particular aspect of one's performance. You do ten kicks with your right leg, then ten kicks with your left. You do ten block-and-punch combinations to the right, then ten to the left. You get into a zone, your mind starts to wander, and pretty soon all of the benefit of the practice dissipates.

This goes back to the basic principle we talked about in the first chapter — the importance of engaging in purposeful practice instead of mindless repetition without any clear plan for getting better. If you want to improve in chess, you don't do it by playing chess; you do it with solitary study of the grandmasters' games. If you want to improve in darts, you don't do it by going to the bar with your friends and letting the loser buy the next round; you do it by spending some time alone working on reproducing your throwing motion exactly from one throw to the next. You improve your control by systematically varying the point on the dartboards that you aim at. If you want to get better at bowling, those Thursday nights with your bowling league team won't do much good. You'll want to spend some quality alley time on your own — ideally, working on difficult pin configurations in which being able to control exactly where the ball goes is essential. And so on.

Remember: if your mind is wandering or you're relaxed and just having fun, you probably won't improve.

A little over a decade ago, a group of Swedish researchers studied two groups of people during and after a singing lesson. Half of the subjects were professional singers, and the other half were amateurs. All had been taking lessons for at least six months. The researchers measured the subjects in a variety of ways — an electrocardiogram, blood samples, visual observations of the singers' facial expressions, and so on — and after the lesson they asked a number of questions that were designed to determine the singers' thought processes during the lesson. All of the singers, both amateur and professional, felt more relaxed and energized after the lesson than before, but only the amateurs reported feeling elated afterward. The singing lesson had made the amateurs, but not the professionals, happy. The reason for the difference lay in how the two groups had approached the lesson. For the amateurs it was a time to express themselves, to sing away their cares, and to feel the pure joy of singing. For the professionals, the lesson was a time to concentrate on such things as vocal technique and breath control in an effort to improve their singing. There was focus but no joy.

This is a key to getting the maximum benefit out of any sort of practice, from private or group lessons to solitary practice and even to games or competitions: whatever you are doing, focus on it.

A graduate student who worked with me at Florida State, Cole Armstrong, described high school golfers developing this sort of focus. At some point around their sophomore years they began to understand what it meant to engage in purposeful practice rather than just practice. In his dissertation, Cole quoted one high school golfer explaining when and how the shift occurred in his approach to practice:

> I can think about a specific point sophomore year. My coach came up to me on the range and said, "Justin, what are you doing?" I was hitting balls and said, "I'm practicing for the tournament." And he said, "No, you're not. I've been watching you, and you're just hitting balls. You're not really doing a routine or anything." So we had a conversation, and, like you said, we started a routine, a practice routine, and from then on I really started to practice where it was a conscious action working towards a specific goal, not just hit balls or putt.

Learning to engage in this way — consciously developing and refining your skills — is one of the most powerful ways to improve the effectiveness of your practice.

The American swimmer Natalie Coughlin once described her own version of this sort of "aha" moment. Over her career she won a total of twelve Olympic medals — an accomplishment that tied her with two others for the most Olympic medals ever won by a female swimmer. While she was always a very good swimmer, she didn't become great until she learned to focus throughout her practice. For much of her early swimming career she would pass the time she spent swimming laps by daydreaming. This is common not just with swimmers but also with runners and all the other types of endurance athletes who spend hours and hours every week putting in the distances needed to build their stamina. Stroke, stroke, stroke, stroke, stroke, stroke, over and over again for hours on end; it's hard not to get bored and just zone out, letting your mind wander far outside the pool. And this is what Coughlin did.

But at some point while she was competing for the University of California, Berkeley, Coughlin realized that she'd been wasting a major opportunity during those hours she spent swimming laps. Instead of letting her mind wander, she could be focusing on her technique, trying to make each stroke as close to perfect as possible. In particular, she could be working on sharpening her mental representations of her stroke — figuring out exactly how her body feels during a "perfect" stroke. Once she had a clear idea of what that ideal stroke felt like, she could notice when she deviated from that ideal — perhaps when she was tired or when she was approaching a turn — and then work on ways to minimize those deviations and keep her strokes as close to ideal as possible.

From then on, Coughlin made a point of staying engaged in what she was doing, using the time spent swimming laps to improve her form. It was only when she began doing this that she really started to see improvement in her times, and the more she focused on her form in her training, the more success she had in her meets. Nor is Coughlin an isolated example. After carrying out an extended study of Olympic swimmers, the researcher Daniel Chambliss concluded that the key to excellence in swimming lay in maintaining close attention to every detail of performance, "each one done correctly, time and again, until excellence in every detail becomes a firmly ingrained habit."

This is the recipe for maximum improvement from your practice. Even in those sports such as bodybuilding or long-distance running, where much of the practice consists of seemingly mindless, repetitive actions, paying attention performing those actions the right way will lead to greater improvement. Researchers who have studied long-distance runners have found that amateurs tend to daydream or think about more pleasant subjects to take their minds off the pain and strain of their running, while elite long-distance runners remain attuned to their bodies so that they can find the optimal pace and make adjustments to maintain the best pace throughout the whole race. In bodybuilding or weightlifting, if you are going to attempt to lift a weight at the maximum of your current ability, you need to prepare before the lift and be completely focused during the lift. Any activity at the limits of your ability will require full concentration and effort. And, of course, in fields where strength and endurance are not so important — intellectual activities, musical performance, art, and so on — there is little point at all to practicing if you don't focus.

Maintaining this sort of focus is hard work, however, even for experts who have been doing it for years. As I noted in chapter 4, the violin students I studied at the Berlin academy found their training so tiring that they would often take a midday nap between their morning and afternoon practice sessions. People who are just learning to focus on their practice won't be able to maintain it for several hours. Instead, they'll need to start out with much shorter sessions and gradually work up.

The advice I offered to Per Holmlöv in this area can be applied to just about anyone who is getting started on deliberate practice: Focus and concentration are crucial, I wrote, so shorter training sessions with clearer goals are the best way to develop new skills faster. It is better to train at 100 percent effort for less time than at 70 percent effort for a longer period. Once you find you can no longer focus effectively, end the session. And make sure you get enough sleep so that you can train with maximum concentration.

Per took my advice. He had arranged to get personal sessions with his sensei, he was doing shorter training sessions but at a higher level of concentration, and he was getting seven to eight hours of sleep a night and a nap after lunch. He had passed his green belt test, and his next goal was the blue belt. At seventy years old he was halfway to black belt, and as long as he stayed injury free, he was confident he would reach that goal before he turned eighty.

IF YOU DON'T HAVE A TEACHER

The last time we encountered Benjamin Franklin in this book, he was playing chess for hours and hours but never really getting any better. This provided us with an excellent example of how *not* to practice — just doing the same thing over and over again without any focused step-by-step plan for improvement. But Franklin was far more than a chess player, of course. He was a scientist, inventor, diplomat, publisher, and a writer whose words are still read more than two centuries later. So let us give equal time to an area in which he did much better than he did in chess.

Early in his autobiography Franklin describes how as a young man he worked to improve his writing. The education he had received as a child had left him, by his own assessment, not much more than an average writer. Then he ran across an issue of the British magazine *The Spectator* and found himself impressed by the quality of the writing in its pages. Franklin decided that he would like to write that well, but he had no one to teach him how. What could he do? He came up with a series of clever techniques aimed at teaching himself how to write as well as the writers of *The Spectator*.

He first set out to see how closely he could reproduce the sentences in an article once he had forgotten their exact wording. So he chose several of the articles whose writing he admired and wrote down short descriptions of the content of each sentence — just enough to remind him what the sentence was about. After several days he tried to reproduce the articles from the hints he had written down. His goal was not so much to produce a word-for-word replica of the articles as to create his own articles that were as detailed and well written as the original. Having written his reproductions, he went back to the original articles, compared them with his own efforts, and corrected his versions where necessary. This taught him to express ideas clearly and cogently.

The biggest problem he discovered from these exercises was that his vocabulary was not nearly as large as those of the writers for *The*

Spectator. It wasn't that he didn't know the words, but rather that he didn't have them at his fingertips when he was writing. To fix this he came up with a variation of his first exercise. He decided that writing poetry would force him to come up with a plethora of different words that he might not normally think of because of the need to fit the poem's rhythm and the rhyming pattern, so he took some of the *Spectator* articles and transformed them into verse. Then, after waiting long enough that his memory of the original wording had faded, he would transform the poems back into prose. This got him into the habit of finding just the right word and increased the number of words he could call up quickly from his memory.

Finally, Franklin worked on the overall structure and logic of his writing. Once again, he worked with articles from *The Spectator* and wrote hints for each sentence. But this time he wrote the hints on separate pieces of paper and then jumbled them so that they were completely out of order. Then he waited long enough that not only had he forgotten the wording of the sentences in the original articles, but he had also forgotten their order, and he tried once again to reproduce the articles. He would take the jumbled hints from one article and arrange them in what he thought was the most logical order, then write sentences from each hint and compare the result with the original article. The exercise forced him to think carefully about how to order the thoughts in a piece of writing. If he found places where he'd failed to order his thoughts as well as the original writer, he would correct his work and try to learn from his mistakes. In his typically humble way, Franklin recalled in his autobiography how he could tell that the practice was having its desired effect: "I sometimes had the pleasure of fancying that, in certain particulars of small import, I had been lucky enough to improve the method or the language, and this encouraged me to think I might possibly in time come to be a tolerable English writer, of which I was extremely ambitious."

Franklin was too modest, of course. He went on to become one of the most admired writers of early America, with *Poor Richard's Alma-*

nack and, later, his autobiography becoming classics of American literature. Franklin solved a problem — wanting to improve, but having no one to teach him how — which many people face from time to time. Maybe you can't afford a teacher, or there is no one easily accessible to teach what you want to learn. Maybe you're interested in improving in some area where there are no experts, or at least no teachers. Whatever the reasons are, it is still possible to improve if you follow some basic principles from deliberate practice — many of which Franklin seems to have intuited on his own.

The hallmark of purposeful or deliberate practice is that you try to do something you cannot do — that takes you out of your comfort zone — and that you practice it over and over again, focusing on exactly how you are doing it, where you are falling short, and how you can get better. Real life — our jobs, our schooling, our hobbies — seldom gives us the opportunity for this sort of focused repetition, so in order to improve, we must manufacture our own opportunities. Franklin did it with his exercises, each focused on a particular facet of writing. Much of what a good teacher or coach will do is to develop such exercises for you, designed specifically to help you improve the particular skill you are focused on at the moment. But without a teacher, you must come up with your own exercises.

Fortunately, we live in a time when it is easy to go to the Internet and find training techniques for most of the common skills that people are interested in and quite a few that are not so common at all. Want to improve your puck-handling skills in hockey? It's on the Internet. Want to be a better writer? On the Internet. To solve a Rubik's Cube really fast? Internet. Of course, you have to be careful about the advice — the Internet offers just about everything except quality control — but you can get some good ideas and tips, try them out, and see what works best for you.

But not everything is on the Internet, and the things that are may not fit exactly what you're trying to do or may not be practical. Some of the most challenging skills to practice, for instance, are those that involve interacting with other people. It's easy enough to sit in your room spinning a Rubik's Cube faster and faster or to go to a driving range and practice hitting with your woods, but what if your skill requires a partner or an audience? Devising an effective way to practice such a skill can require some creativity.

Another professor at Florida State University, who worked with ESL (English as a second language) students, told me about a student of his who went to the mall and stopped a number of shoppers, asking each the same question. In this way she was able to hear similar answers over and over again, and that repetition made it easier for her to understand the words being spoken by native speakers at full speed. If she had asked different questions each time, it's likely that her comprehension would have improved little, if at all. Other students who were trying to improve their English would watch the same English movies with subtitles over and over again, covering the subtitles and trying to understand what was being said. To check their comprehension, they would uncover the subtitles. By listening to the same dialogue over and over, they improved their ability to understand English much more quickly than if they'd simply watched a number of different movies.

Note that these students weren't simply doing the same thing over and over again: they were paying attention to what they got wrong each time and correcting it. This is purposeful practice. It does no good to do the same thing over and over again mindlessly; the purpose of the repetition is to figure out where your weaknesses are and focus on getting better in those areas, trying different methods to improve until you find something that works.

One of my favorite examples of this sort of clever self-designed practice technique was described to me by a student at a circus school in Rio de Janeiro. He was training to be a ringmaster, and his problem was how to keep the audience interested during a show. Besides introducing the various circus acts, the ringmaster must be ready to fill up any empty time between acts if there is some sort of delay in presenting the next act. But no one was going to let this student practice his technique with live audiences, so he came up with an idea. He went to downtown Rio and struck up conversations with people who were going home during rush hour. Most of them were in a hurry, so he had to work to keep them interested enough to stay and listen to what he had to say. In doing so he got to practice using his voice and body language to draw attention to himself and using pauses that were long enough, but not too long, to create dramatic tension.

What struck me most, though, was how deliberate he was about it: He used his watch to time exactly how long he could keep each conversation going. He spent a couple of hours each day doing this, taking notes about which techniques worked best and which didn't work well at all.

Comedians do something very similar. There is a reason that most of them have spent time in standup comedy clubs. They get a chance to try out their material and their delivery, and they get immediate feedback from the audience: either the jokes work, or they don't. And they can come back night after night, honing their material, getting rid of what doesn't work and making what does work even better. Even established comedians will often return to standup clubs to try out new routines or simply brush up on their delivery.

To effectively practice a skill without a teacher, it helps to keep in mind three Fs: Focus. Feedback. Fix it. Break the skill down into components that you can do repeatedly and analyze effectively, determine your weaknesses, and figure out ways to address them.

The ringmaster, the ESL students, and Ben Franklin all exemplified this approach. Franklin's approach also offers an excellent template for developing mental representations when you have little or no input from instructors. When he analyzed the writing in *The Spectator* and figured out what made it good, he was — although he didn't think of it in these terms — creating a mental representation that he could use to guide his own work. The more he practiced, the more highly developed his mental representations became, until he could write at the level of *The Spectator* without having a concrete example in front of him. He had internalized good writing—which is just another way of saying that he had built mental representations that captured its salient features.

Ironically, this is exactly what Franklin failed to do as a chess player. With writing, he studied the work of experts and tried to reproduce it; when he failed to reproduce it well enough, he would take another look at it and figure out what he had missed so that he would do better the next time. But this is exactly how chess players improve most effectively—by studying the games of grandmasters, trying to reproduce them move by move, and, when they choose a move that is different from what the grandmaster chose, studying the position again to see what they missed. Franklin could not apply this technique to chess, however, because he had no easy access to the games of masters. Almost all of them were in Europe, and at the time there were no books with their collected games for him to study. If he had had some way to study the masters' games, he might well have become one of the best chess players of his generation. He was certainly one of its best writers.

We can build effective mental representations in many areas with a similar technique. In music, Wolfgang Amadeus Mozart's father taught him to compose in part by having him study some of the era's best composers and copy their work. And in art, aspiring artists have long developed their skills by copying the paintings and sculptures of the masters. Indeed, in some cases they have done this in a way very similar to the technique Franklin used to improve his writing, that is, by studying a piece of art by a master, attempting to reproduce it from memory, and then comparing the finished product with the original in order to discover the differences and correct them. Some artists even become so good at copying that they can make their living as forgers, but that is not usually the point of this exercise. Artists don't want to produce artwork that looks like someone else's; they want to develop the skills and the mental representations that make expertise possible and use that expertise to convey their own artistic vision.

Despite the first word in the term "mental representation," pure *mental* analysis is not nearly enough. We can only form effective mental representations when we try to reproduce what the expert performer can do, fail, figure out why we failed, try again, and repeat — over and over again. Successful mental representations are inextricably tied to actions, not just thoughts, and it is the extended practice aimed at reproducing the original product that will produce the mental representations we seek.

GETTING PAST PLATEAUS

In 2005 a young journalist named Joshua Foer came to Tallahassee to interview me about an article that he was writing about memory competitions. These are the sorts of events that I mentioned earlier, where people compete to see who can recall the most digits, who can memorize a random collection of playing cards most quickly, and other similar feats. During our discussions Josh mentioned that he was thinking about competing himself in order to get a first-person perspective and that he was going to start training under a top-ranked memory competitor, Ed Cooke. There was even some vague talk about a book he might write on his experiences in these competitions.

Before Josh began working with Cooke, my graduate students and I tested his memory on a wide variety of tasks to see what his baseline abilities were. After that we had little contact for a while, until one day he called me and complained that he'd reached a plateau. No matter how much he practiced, he couldn't improve the speed with which he memorized the order of a randomly arranged deck of cards.

I gave Josh some advice about getting past a plateau, and he went back to training. The whole story is told in his book *Moonwalking with Einstein,* but the bottom line is this: Josh did indeed speed up considerably, and he ultimately won the 2006 USA Memory Championship.

The plateau Josh encountered is common in every sort of training. When you first start learning something new, it is normal to see rapid—or at least steady—improvement, and when that improvement stops, it is natural to believe you've hit some sort of implacable limit. So you stop trying to move forward, and you settle down to life on that plateau. This is the major reason that people in every area stop improving.

I had run into this very problem in my work with Steve Faloon. Steve had been stuck at about the same number of digits for several weeks and thought that he might have reached his limit. Since he was already beyond what anyone else had ever done, Bill Chase and I didn't know what to expect. Had Steve gone as far as one could possibly go? And how would we even know if he had reached some upper limit? We decided to do a little experiment. I slowed down the rate at which I read out the digits. It was just a minor adjustment, but it gave Steve enough extra time to hold on to significantly more digits than he'd ever managed before. This convinced him that the problem was not the number of digits but rather how quickly he was encoding the numbers. He believed that he might improve his performance if he could just speed up the time it took him to commit the digits to long-term memory.

At another plateau Steve found that he was consistently messing up a couple of digits in one of his digit groups when he was given strings of a certain length. He worried he might have reached his limit on how many digit groups he could recall correctly. So Bill and I gave him strings that were ten or more digits longer than he had ever managed to remember. He surprised himself by remembering most of the digits—and, in particular, remembering more total digits than he'd ever done before, even though he wasn't perfect. This showed him that it was indeed possible to remember longer strings of digits and that his problem was not that he had reached the limit of his memory but rather that he was messing up on one or two groups of digits in the entire string. He started focusing on encoding the digit groups more carefully in his long-term memory, and he rose above that plateau as well.

What we learned from Steve's experience holds true for everyone who faces a plateau: the best way to move beyond it is to challenge your brain or your body in a new way. Bodybuilders, for instance, will change the types of exercises they are doing, increase or decrease the weight they're lifting or the number of repetitions, and switch up their weekly routine. Actually, most of them will vary their patterns proactively so they don't get stuck on plateaus in the first place. Cross-training of any sort is based on the same principle — switch off between different types of exercise so that you are constantly challenging yourself in different ways.

But sometimes you try everything you can think of and you're still stuck. When Josh came to me for help with his card memorization, I told him about what had worked with Steve, and we talked about why.

We also talked about typing. People who learn to type with the classic ten-finger method, where each finger is assigned to certain keys, will eventually reach a certain comfortable speed at which they can type maybe thirty or forty words per minute with relatively few mistakes. This is their plateau.

Typing teachers use a well-established method to get past such a plateau. Most typists can increase their typing speed by 10–20 percent simply by focusing and pushing themselves to type faster. The problem is that as their concentration lags, their typing speed returns to the plateau. To counter this, a teacher will typically suggest setting aside fifteen to twenty minutes a day to type at this faster speed.

This does two things: First, it helps the student spot challenges — such as particular letter combinations — that slow down their typing. Once you figure out what the problems are, you can design exercises to improve your speed in those situations. For instance, if you're having a problem typing "ol" or "lo" because the letter *o* is almost directly above the letter *l*, you could practice typing a series of words that contain those combinations — old, cold, roll, toll, low, lot, lob, lox, follow, hollow, and so on — over and over again.

Second, when you type faster than usual, it forces you to start looking ahead at the words that are coming up so that you can figure out where to place your fingers in anticipation. So, if you see that the next four letters will all be typed by fingers on your left hand, you can move the correct finger on your right hand into place for the fifth letter ahead. Tests on the best typists have shown that their speeds are closely related to how far ahead they look at upcoming letters while they type.

Although both typing and digit memorization are very specialized skills, the methods of getting past a plateau used in the two areas point toward an effective general approach to plateaus. Any reasonably complex skill will involve a variety of components, some of which you will be better at than others. Thus, when you reach a point at which you are having difficulty getting better, it will be just one or two of the components of that skill, not all of them, that are holding you back. The question is, Which ones?

To figure that out, you need to find a way to push yourself a little — not a lot — harder than usual. This will often help you figure out where your sticking points are. If you're a tennis player, try playing a better opponent than you are used to; your weaknesses will probably become much more obvious. If you're a manager, pay attention to what goes wrong when things get busy or chaotic — those problems are not anomalies but rather indications of weaknesses that were probably there all the time but were usually less obvious.

With all of this in mind, I suggested to Josh that if he wanted to speed up the pace at which he could memorize the order of a deck of cards, he should try to do it in less time than it normally took and then look to see where his mistakes were coming from. By identifying exactly what was slowing him down, he could come up with exercises to improve his speed on those particular things instead of simply trying, over and over again, to produce some generalized improvement that would decrease the amount of time he spent on an entire deck of cards.

This, then, is what you should try when other techniques for getting past a plateau have failed. First, figure out exactly what is holding you back. What mistakes are you making, and when? Push yourself well outside of your comfort zone and see what breaks down first. Then design a practice technique aimed at improving that particular weakness. Once you've figured out what the problem is, you may be able to fix it yourself, or you may need to go to an experienced coach or teacher for suggestions. Either way, pay attention to what happens when you practice; if you are not improving, you will need to try something else.

The power of this technique is that it targets those specific problem areas that are holding you back rather than trying this and that and hoping that something works. This technique is not widely recognized, even among experienced teachers, even though it might seem obvious as described here and is a remarkably effective way to rise above plateaus.

MAINTAINING MOTIVATION

In the summer of 2006, 274 middle-schoolers traveled to Washington, D.C., for the Scripps National Spelling Bee, which would eventually be won by Kerry Close, a thirteen-year-old from Spring Lake, New Jersey, with the word *ursprache* in the twentieth round. My students and I were there to find what set apart the very best spellers from the rest.

We gave each contestant a detailed questionnaire asking about their study practices. The questionnaires also included items designed to assess the contestants' personalities. Spelling contestants have two basic approaches to preparing for a contest — spending time alone studying words from various lists and dictionaries, and being quizzed by others on words from those lists. We found that when the contestants first started out, they generally spent more time being quizzed by others, but later they relied more on solitary practice. When we compared how well the various contestants did in the bee versus their study histories, we found that the top spellers had spent significantly more time than their peers in purposeful practice — mainly, solo sessions in which they focused on memorizing the spelling of as many words as possible. The best spellers had also spent more time on being quizzed, but the amount of time they spent in purposeful practice correlated more closely with how well they did in the spelling bee.

What we were really interested in, however, was what motivated these students to spend so much time studying the spelling of words. The students who win the regional competitions and go on to compete in the national spelling bee — even those who don't end up among the top spellers at the event — put in incredible amounts of practice time in the months before the competition. Why? In particular, what drove the very best spellers to put in so much more time than the others?

Some people had suggested that the students who had spent the most time practicing did so because they actually liked this sort of studying and got some sort of pleasure out of it. But the answers the students gave to our questionnaire told a very different story: they didn't like studying at all. None of them did, including the very best spellers. The hours they had spent studying thousands of words alone were not fun; they would have been quite happy to do something else. Instead, what distinguished the most successful spellers was their superior ability to remain committed to studying despite the boredom and the pull of other, more appealing activities.

How do you keep going? That is perhaps the biggest question that anyone engaged in purposeful or deliberate practice will eventually face.

Getting started is easy, as anyone who has visited a gym after New Year's knows. You decide that you want to get in shape or learn to play the guitar or pick up a new language, and so you jump right in. It's exciting. It's energizing. You can imagine how good it will feel to be twenty pounds lighter or to play "Smells Like Teen Spirit." Then after a while, reality hits. It's hard to find the time to work out or practice as much as you should, so you start missing sessions. You're not improving as fast as you thought you would. It stops being fun, and your resolve to reach your goal weakens. Eventually you stop altogether, and you don't start up again. Call it "the New Year's resolution effect"—it's why gyms that were crowded in January are only half full in July and why so many slightly used guitars are available on Craigslist.

So that's the problem in a nutshell: purposeful practice is hard work. It's hard to keep going, and even if you keep up your training—you go to the gym regularly, or you practice the guitar for a certain number of hours every week—it's hard to maintain focus and effort, so you may eventually stop pushing yourself and stop improving. The question is, What can you do about it?

In answering that question the first thing to note is that, despite the effort that it takes, it certainly is possible to keep going. Every world-class athlete, every prima ballerina, every concert violinist, every chess grandmaster is living proof that it can be done—that people can practice hard day after day, week after week, for years on end. These people have all figured out how to get past the New Year's resolution effect and make deliberate practice an ongoing part of their lives. How did they do it? What can we learn from expert performers about what it takes to keep going?

Let's get one thing out of the way right up front. It may seem natural to assume that these people who maintain intense practice schedules for years have some rare gift of willpower or "grit" or "stick-to-itiveness" that the rest of us just lack, but that would be a mistake for two very compelling reasons.

First, there is little scientific evidence for the existence of a general "willpower" that can be applied in any situation. There is no indication, for example, that the students who had enough "willpower" to study countless hours for the national spelling bee would show the same amount of "willpower" if they were asked to practice the piano or chess or baseball. In fact, if anything, the available evidence indicates that willpower is a very situation-specific attribute. People generally find it much easier to push themselves in some areas than in others. If Katie became a grandmaster after ten years of studying chess and Karl gave up on the game after six months, does that mean Katie had more willpower than Karl? Would it change your answer if I told you that Katie spent a year practicing the piano and then quit before she started chess, while Karl is now an internationally renowned concert pianist? This situational dependence calls into question the claim that some sort of generic willpower can explain an individual's ability to sustain daily practice for months, years, and decades.

But there is a bigger, second problem with the concept of *willpower,* one related to the myth of natural talent, which we will discuss later in chapter 8. Both willpower and natural talent are traits that people assign to someone after the fact: Jason is an incredible tennis player, so he must have been born with this natural talent. Jackie practiced the violin for years, several hours each day, so she must have incredible willpower. In neither case can we make this determination ahead of time with any likelihood of being right, and in neither case has anyone ever identified any genes that underlie these supposed innate characteristics, so there is no more scientific evidence for the existence of individual genes that determine willpower than there is for the existence of genes that are necessary for succeeding in chess or piano-playing. Furthermore, once you assume that something is innate, it automatically becomes something you can't do anything about: If you don't have innate musical talent, forget about ever being a good musician. If you don't have enough willpower, forget about ever taking on something that will require a great deal of hard work. This sort of circular thinking — "The fact that I couldn't keep practicing indicates that I don't have enough willpower, which explains why I couldn't keep practicing" — is worse than useless; it is damaging in that it can convince people that they might as well not even try.

It is much more useful, I believe, to talk about motivation. Motivation is quite different from willpower. We all have various motivations — some stronger, some weaker — at various times and in various situations. The most important question to answer then becomes, What factors shape motivation? By asking such a question, we can home in on the factors that might boost the motivation of our employees, children, students, and ourselves.

There are some interesting parallels between improving performance and losing weight. People who are overweight generally have little difficulty starting a diet program, and they will generally lose some weight on it. But almost all of them will eventually see their progress stop, and most of those will gradually regain the weight they lost, putting them right back where they started. The ones who are successful in losing weight over the long run are those who have successfully redesigned their lives, building new habits that allow them to maintain the behaviors that keep them losing weight in spite of all of the temptations that threaten their success.

A similar thing is true for those who maintain purposeful or deliberate practice over the long run. They have generally developed various habits that help them keep going. As a rule of thumb, I think that anyone who hopes to improve skill in a particular area should devote an hour or more each day to practice that can be done with full concentration. Maintaining the motivation that enables such a regimen has two parts: reasons to keep going and reasons to stop. When you quit something that you had initially wanted to do, it's because the reasons to stop eventually came to outweigh the reasons to continue. Thus, to maintain your motivation you can either strengthen the reasons to keep going or weaken the reasons to quit. Successful motivation efforts generally include both.

There are various ways to weaken the reasons to quit. One of the most effective is to set aside a fixed time to practice that has been cleared of all other obligations and distractions. It can be difficult enough to push yourself to practice in the best of situations, but when you have other things you could be doing, there is a constant temptation to do something else and to justify it by telling yourself that it really needs to get done. If you do this often enough, you begin practicing less and less, and soon your training program is in a death spiral.

When I studied the violin students in Berlin I found that most of them preferred to practice as soon as they got up in the morning. They had set up their schedules so that there was nothing else to do at that time. It was set aside specifically for practice. Furthermore, identifying that period as their practice time created a sense of habit and duty that made it less likely they'd be tempted by something else. The best and the better students averaged around five hours more of sleep per week than the good students, mostly by taking more time for afternoon naps. All of the students in the study — the good students, the better, and the best — spent about the same amount of time each week on leisure activities, but the best students were much better at estimating how much time they spent on leisure, which indicates that they made more of an effort to plan their time. Good planning can help you avoid many of the things that might lead you to spend less time on practice than you wanted.

More generally, look for anything that might interfere with your training and find ways to minimize its influence. If you're likely to be distracted by your smartphone, turn it off. Or better yet, turn it off and leave it in another room. If you're not a morning person and you find it particularly difficult to exercise in the morning, move your run or your exercise class to later in the day when your body won't fight you so much. I've noticed that some people who have difficulty getting started in the morning don't get enough sleep. Ideally you should wake up by yourself (that is, without an alarm to wake you) and feel refreshed when you do. If that's not the case, you might need to go to bed earlier. While any given factor may make only a small difference, the various factors add up.

For purposeful or deliberate practice to be effective, you need to push yourself outside your comfort zone and maintain your focus, but those are mentally draining activities. Expert performers do two things — both seemingly unrelated to motivation — that can help. The first is general physical maintenance: getting enough sleep and keeping healthy. If you're tired or sick, it's that much harder to maintain focus and that much easier to slack off. As I mentioned in chapter 4, the violin students were all careful to get a good night's sleep each night, and many of them would take an early afternoon nap after their morning practice session. The second thing is to limit the length of your practice sessions to about an hour. You can't maintain intense concentration for much longer than that — and when you're first starting out, it's likely to be less. If you want to practice longer than an hour, go for an hour and take a break.

Fortunately, you will find that as you maintain your practice over time it will seem easier. Both your body and your mind will habituate to the practice. Runners and other athletes find that they become inured to the pain associated with their exercise. Interestingly, studies have found that while athletes get acclimated to the particular type of pain associated with their sport, they do not get acclimated to pain in general. They still feel other types of pain just as acutely as anyone else does. Similarly, over time musicians and anyone else who practices intensely get to the point where those hours of practice no longer seem as mentally painful as they once were. The practice never becomes outright fun, but eventually it gets closer to neutral, so it's not as hard to keep going.

We've just seen several ways to decrease the inclination to stop; now let's look at some ways to increase the inclination to continue.

The motivation must, of course, be a desire to be better at whatever it is you are practicing. If you don't have that desire, why are you practicing? But that desire may come in different forms. It may be completely intrinsic. Say you've always wanted to be able to make origami figures. You don't know why, but it's inside you. Sometimes the desire is part of something larger. You love listening to the symphony, and you've decided that you would really like to be part of that — a member of an orchestra contributing to that amazing sound and experiencing it from that perspective — but you don't have an overriding desire to play the clarinet or the saxophone or any other particular instrument. Or maybe it's for totally practical, extrinsic purposes. You hate public speaking, but you recognize that your lack of speaking skills is holding you back in your career, so you decide you want to learn how to address an audience. All of these are possible roots of motivation, but they aren't — or at least they shouldn't be — your only motivators.

Studies of expert performers tell us that once you have practiced for a while and can see the results, the skill itself can become part of your motivation. You take pride in what you do, you get pleasure from your friends' compliments, and your sense of identity changes. You begin to see yourself as a public speaker or a piccolo player or a maker of origami figures. As long as you recognize this new identity as flowing from the many hours of practice that you devoted to developing your skill, further practice comes to feel more like an investment than an expense.

Another key motivational factor in deliberate practice is a belief that you can succeed. In order to push yourself when you really don't feel like it, you must believe that you can improve and — particularly for people shooting to become expert performers — that you can rank among the best. The power of such belief is so strong that it can even trump reality. One of Sweden's most famous athletes, the middle-distance runner Gunder Hägg, who broke fifteen world records in the early 1940s, grew up with his father, a lumberjack, in an isolated part of northern Sweden. In his early teen years Gunder loved running in the woods, and he and his father became curious about how fast he could run. They found a route that was about fifteen hundred meters long, and Gunder ran that course while his father measured his time with an alarm clock. When Gunder was done, his father told him that he finished the distance in 4 minutes, 50 seconds — a remarkably good time for that distance in the woods. As he would later recall in his autobiography, Gunder was inspired by his performance to believe he had a bright future as a runner, so he started training more seriously, and indeed he did go on to become one of the world's premier runners. It was only many years later that his father confessed to him that the actual time on that day was 5 minutes, 50 seconds and that he had exaggerated Gunder's speed because he was worried that Gunder had lost some of his passion for running and needed to be encouraged.

The psychologist Benjamin Bloom once directed a project that examined the childhoods of a number of experts in various fields. One of his findings was that when these future experts were young, their parents would use various strategies to keep them from quitting. In particular, several of the experts told of a time in their youth when they were sick or injured in some way that prevented them from practicing for a significant amount of time. When they eventually resumed practicing, they weren't at nearly the same level they had been at before, and, discouraged, they wanted to quit. Their parents told them that they could quit if they wished but that first they needed to keep practicing enough to get back to where they were. And this did the trick. Once they'd practiced for a while and gotten back to where they were, they realized that they could indeed keep getting better and that their setback was just temporary.

Belief is important. You may not be lucky enough to have someone do for you what Hägg's father did for him, but you can certainly take a lesson from the expert performers that Bloom studied: if you stop believing that you can reach a goal, either because you've regressed or you've plateaued, don't quit. Make an agreement with yourself that you will do what it takes to get back to where you were or to get beyond the plateau, and then you can quit. You probably won't.

One of the strongest forms of extrinsic motivation is social motivation. This can take several forms. One of the simplest and most direct is the approval and admiration of others. Young children are often motivated to practice a musical instrument or a sport because they are looking for their parents' approval. Older children, on the other hand, are often motivated by positive feedback for their accomplishments. After having practiced long enough to reach a certain skill level, they become known for their abilities — this child is an artist, that child plays the piano well, and that one is a phenomenal basketball player — and this recognition can provide motivation to keep going. Many teenagers — and more than a few adults — have taken up a musical instrument or a sport because they believed that expertise in that area would make them more sexually attractive.

One of the best ways to create and sustain social motivation is to surround yourself with people who will encourage and support and challenge you in your endeavors. Not only did the Berlin violin students spend most of their time with other music students, but they also tended to date music students or at least others who would appreciate their passion for music and understand their need to prioritize their practice.

Surrounding yourself with supportive people is easiest in activities that are done in groups or teams. If you're part of an orchestra, for example, you may find yourself motivated to practice harder because you don't want to let your colleagues down, or because you're competing with some of them to be the best at your instrument, or perhaps both. The members of a baseball or softball team may collectively push to improve in order to win a championship, but they will also be aware of the internal competitions with other members of the team and will likely be motivated by those competitions as well.

Perhaps the most important factor here, though, is the social environment itself. Deliberate practice can be a lonely pursuit, but if you have a group of friends who are in the same positions — the other members of your orchestra or your baseball team or your chess club — you have a built-in support system. These people understand the effort you're putting into your practice, they can share training tips with you, and they can appreciate your victories and commiserate with you over your difficulties. They count on you, and you can count on them.

I asked Per Holmlöv what would motivate a seventy-something-year-old man to devote many hours every week to earning a black belt. He told me that he first became interested in karate because his grandchildren had started training, and he enjoyed watching them and interacting with them as they trained. But what drove his training over the years was his interaction with his fellow students and his teachers. Much of karate training is done in pairs, and Per explained that he'd found a training partner — a woman about twenty-five years younger whose children were also training in the school — who was exceptionally supportive of him and of his advancement in karate. Several young male students in his school were also supportive, and these compatriots provided his strongest motivation to keep going.

In my most recent communications with Per — in the summer of 2015, when he was seventy-four — I learned that he and his wife had moved near the mountains to Åre, the Swedish equivalent of Aspen, Colorado. He had reached the level of blue belt and had been planning to test for his brown belt, but since he no longer had the opportunity to train in a karate school with other students, he had decided that he had to give up on his advancement toward black belt. He still trains each morning with a routine his sensei developed for him, a routine that includes warming up, karate forms, work with kettle bells, and meditation, and he regularly hikes in the mountains. His life goals, he wrote me, are "wisdom and vitality."

Which brings us back to Benjamin Franklin again. As a young man he was interested in all sorts of intellectual pursuits — philosophy, science, invention, writing, the arts, and so on — and he wished to encourage his own development in those areas. So at twenty-one he recruited eleven of the most intellectually interesting people in Philadelphia to form a mutual improvement club, which he named "the Junto." The club's members, who met each Friday night, would encourage each other's various intellectual pursuits. Every member was expected to bring at least one interesting topic of conversation — on morals, politics, or science — to each meeting. The topics, which were generally phrased as questions, were to be discussed by the group "in the sincere spirit of inquiry after truth, without fondness for dispute or desire of victory." In order to keep the discussions open and collaborative, the Junto's rules strictly forbade anyone from contradicting another member or expressing an opinion too strongly. And once every three months each member of the Junto had to compose an essay — on any topic whatsoever — and read it to the rest of the group, which would then discuss it.

One purpose of the club was to encourage the members to engage with the intellectual topics of the day. By creating the club Franklin not only ensured himself regular access to some of the most interesting people in the city, but he was giving himself extra motivation (as if he needed any) to delve into these topics himself. Knowing that he was expected to ask at least one interesting question each week and that he would also be answering others' questions gave him extra impetus to read and examine the most urgent and intellectually challenging matters in contemporary science, politics, and philosophy.

This technique can be used in nearly any area: put together a group of people all interested in the same thing — or join an existing group — and use the group's camaraderie and shared goals as extra motivation in reaching your own goals. This is the idea behind many social organizations, from book clubs to chess clubs to community theaters, and joining — or, if necessary, forming — such a group can be a tremendous way for adults to maintain motivation. One thing to be careful about, however, is to make sure that the other members of the group have similar goals for improvement. If you join a bowling team because you are trying to improve your bowling scores and the rest of the team is mainly interested in having a good time, with little concern about whether they win the league title, you're going to be frustrated, not motivated. If you're a guitarist looking to improve enough to make a career out of music, don't join a band whose members just want to get together in someone's garage on Saturday nights and jam. (But do keep in mind that Junto would be a really good name for a rock band.)

Of course, at its core, deliberate practice is a lonely pursuit. While you may collect a group of like-minded individuals for support and encouragement, still much of your improvement will depend on practice you do on your own. How do you maintain motivation for hour after hour of such focused practice?

One of the best bits of advice is to set things up so that you are constantly seeing concrete signs of improvement, even if it is not always major improvement. Break your long journey into a manageable series of goals and focus on them one at a time — perhaps even giving yourself a small reward each time you reach a goal. Piano teachers know, for example, that it is best to break down long-term targets for a young piano student into a series of levels. By doing this, the student gets a sense of achievement each time a new level is attained, and that sense of achievement will both add to his or her motivation and make it less likely that the student will become discouraged by a seeming lack of progress. It doesn't matter if the levels are arbitrary. What matters is that the teacher divides up what can look like an infinite amount of material to learn into a series of clear steps, making the student's progress more concrete and more encouraging.

Dan McLaughlin — the golfer of the Dan Plan — has done something very similar in his quest to reach the PGA Tour. From the start, he broke his quest into a series of stages, each devoted to a particular technique, and at each step he developed ways to monitor his progress so that he knew where he was and how far he'd come. Dan's first step was learning how to putt, and for several months the putter was the only golf club he handled. He created various games in which he would repeat the same attempts over and over again, and he kept close track of how he did in these games. In one early game, for example, he would mark off six spots that were each three feet from a hole and distributed evenly around the hole. Then he would try to sink the ball into the hole with putts from each of these six spots and repeat that seventeen times for a total of 102 putts. For each set of 6 putts, Dan counted how many times he got the ball in the hole, and he recorded the scores on a spreadsheet. In this way he could watch his progress in a very concrete way. Not only was he able to tell what sorts of errors he was making and what he needed work on, but he could see, week by week, just how far he had progressed.

Later, after Dan had learned to use the other golf clubs one by one — first a pitching wedge, then the irons, the woods, and finally the driver — he would play his first full round of golf with a complete set of clubs in December 2011, more than a year and a half after he had started, and by this time he was recording his progress in several different ways. He kept track of his driving accuracy, how often his shots that were hit off the tee landed in the fairway, how often they missed to the right, and how often they missed to the left. He kept track of the average number of putts it took to get the ball in the hole once he'd gotten it onto the green. And so on. Not only did the numbers let him see which areas needed work and what sort of work was needed, they served as the mileage markers on his road to golf expertise.

As anyone familiar with golf knows, the most important indicator of Dan's progress is his handicap. The formula for calculating the handicap is somewhat complicated, but in essence it tells you how good a game Dan could be expected to play on one of his better days. Someone with a handicap of 10, for example, is assumed to be able to play eighteen holes of golf at ten strokes over par. The handicap makes it possible for players of different ability levels to play on something approaching an even footing. And because one's handicap is based on scores over the previous twenty or so full rounds of golf one has played, it is constantly changing and provides a record of how well a person has been playing over time.

When Dan first started calculating and recording his handicap, in May 2012, it was 8.7, which was quite good for someone who'd been playing just a couple of years. By the second half of 2014 his handicap was fluctuating between 3 and 4, which was truly impressive. At this writing, in the second half of 2015, Dan was recovering from an injury that had set him back and kept him from playing for a while. He had put in more than six thousand hours of practice, so he was more than 60 percent of the way to his goal of ten thousand hours of practice.

We still don't know if Dan will achieve his goal of playing on the PGA Tour, but he has clearly shown how a thirty-year-old man with no real golf experience can, with the right sort of practice, turn himself into an expert golfer.

My inbox is full of stories like this. A psychotherapist from Denmark who used deliberate practice to develop her singing and eventually recorded songs that have gotten airplay on radio stations all over Denmark. A mechanical engineer from Florida who developed his painting skills through deliberate practice and sent me a picture of his first-ever painting, which was really quite good. A Brazilian engineer who decided to devote ten thousand hours (that number again!) to becoming an origami expert. And so on. The only two things that all these people have in common are that they all have had a dream and that they've all realized, after learning about deliberate practice, that there is a path to achieving that dream.

And this, more than anything else, is the lesson that people should take away from all these stories and all this research: There is no reason not to follow your dream. Deliberate practice can open the door to a world of possibilities that you may have been convinced were out of reach. Open that door.

7

The Road to Extraordinary

IN THE LATE 1960S the Hungarian psychologist László Polgár and his wife, Klara, embarked on a grand experiment that would consume their lives for the next quarter century. László had studied hundreds of people who were considered geniuses in one field or another, and he'd concluded that with the proper rearing any child could be turned into a genius. When he was wooing Klara, he outlined his theories and explained that he was looking for a wife who would collaborate with him to test his theories on their own children. Klara, a teacher from Ukraine, must have been a very special woman, for she responded positively to this unorthodox courtship and agreed to László's proposals (for marriage and for turning their future children into geniuses).

László was so sure his training program would work for any area that he wasn't picky about which particular one he and Klara would target, and the two of them discussed various options. Languages were one option: Just how many languages might it be possible to teach a child? Mathematics was another possibility. Top-flight mathematicians were highly regarded in Eastern Europe at the time, as the Communist regimes sought ways to prove their superiority over the decadent West. Mathematics would have the added advantage that there were no top-level female mathematicians at the time, so, assuming that he and Klara had a daughter, László would prove his claim even more convincingly. But he and Klara settled on a third option.

"We could do the same thing with any subject, if you start early, spend lots of time and give great love to that one subject," Klara would later tell a newspaper reporter. "But we chose chess. Chess is very objective and easy to measure."

Chess had always been thought of as a game for the "male mind," with female chess players treated as second-class citizens. The women had their own tournaments and championships because it was thought it wouldn't be fair to put them up against the men, and there had never been a female grandmaster. Indeed, at the time, the common attitude toward women playing chess was much like Samuel Johnson's famous quote: "A woman's preaching is like a dog's walking on his hind legs. It is not done well; but you are surprised to find it done at all."

The Polgárs were blessed with three children, all of them girls. All the better to prove László's point.

Their first daughter, born in April 1969, was named Susan (in Hungarian, Zsuzsanna). Sofia (Zsófia) followed in November 1974, and then Judit in July 1976. László and Klara home-schooled their daughters in order to spend as much time as possible focusing on chess. It didn't take long for the Polgárs' experiment to become a tremendous success.

Susan was just four years old when she won her first tournament, dominating the Budapest Girls' Under-11 Championship with ten wins, no losses, and no ties. At fifteen, she became the top-ranked woman chess player in the world, and she went on to become the first woman to be awarded grandmaster status via the same path that the males must take. (Two other women had been named grandmasters after winning female-only world championships.) And Susan wouldn't even be the most accomplished of the girls.

Sofia, the second daughter, also had an amazing chess career. Perhaps its highlight came when she was only fourteen, when she dominated a tournament in Rome that included several highly regarded male grandmasters. By winning eight of her nine games and drawing the ninth, she earned a single-tournament chess rating — that is, a rating based only on the games of that tournament — of 2735, which was one of the highest tournament ratings ever for either a male or female player. That was in 1989, and people in the chess world still talk about "the sack of Rome." Although Sofia's highest overall chess rating was 2540, well over the 2500 threshold for grandmaster, and although she had performed more than well enough in sanctioned tournaments, she was never awarded grandmaster status — a result that was apparently more a political decision than a judgment about her chess prowess. (Like her sisters, she never tried to make nice with the male chess establishment.) Sofia was at one time the sixth-ranked female chess player in the world. Among the Polgár sisters, though, she could be considered the slacker.

Judit was the crown jewel of László Polgár's experiment. She became a grandmaster at fifteen years, five months, making her at that time the youngest person, male or female, to ever reach that level. She was the number-one-ranked women's chess player in the world for twenty-five years, until she retired from chess in 2014. At one time she was ranked number eight in the world among all chess players, male or female, and in 2005 she became the first — and so far only — woman to play in the overall World Chess Championship.

The Polgár sisters were all clearly experts. Each of them became among the very best in the world in an area in which the measured performance is extremely objective. There are no style points in chess. Your school background doesn't matter. Your résumé doesn't count. So we know without any doubts just how good they were, and they were very, very good.

And while some details of their background are unusual — very few parents are so focused on turning their children into the best in the world at something — they provide a clear, if somewhat extreme, example of what it takes to become an expert performer. The paths that Susan, Sofia, and Judit took to chess mastery are in line with the path that essentially all experts have taken to become extraordinary. In particular, psychologists have found that an expert's development passes through four distinct stages, from the first glimmers of interest to full-fledged expertise. Everything we know about the Polgár sisters suggests that they went through those same stages, if perhaps in a slightly different fashion because of how their father directed their development.

In this chapter we take an in-depth look at what it takes to become an expert performer. As I explained earlier, most of what we know about deliberate practice has come from the study of experts and how they develop their extraordinary abilities, but to this point in the book we have focused mainly on what this all means for the rest of us — those of us who may use the principles of deliberate practice to get better but who may never be among the best in the world at what we do. Now we switch our attention to those best-in-the-world sorts — the world-class musicians, the Olympic athletes, the Nobel Prize–winning scientists, the chess grandmasters, and the rest.

In one sense this chapter could be thought of as a how-to manual for creating an expert — a road map to excellence, if you will. This chapter won't give you everything you'll need to produce the next Judit Polgár or Serena Williams, but you will leave it with a much better idea of what you're signing on to, if that's the route you choose.

More broadly, this chapter provides a step-by-step look at what is required to take full advantage of human adaptability and reach the frontier of human capabilities. Typically that process begins in childhood or early adolescence and proceeds for a decade or more until the expert level is reached. But it doesn't stop there. One of the hallmarks of expert performers is that even once they become one of the best at what they do, they still constantly strive to improve their practice techniques and to get better. And it is here at the frontier that we find the pathbreakers, those experts who go beyond what anyone else has ever done and show us all what it is possible to achieve.

STARTING OUT

In a magazine interview Susan Polgár spoke about how she first got interested in chess. "I found my first chess set when I was looking in the closet at home for a new toy," she said. "I originally was attracted to the shape of the figures. Later, it was the logic that fascinated me and the challenge."

It is interesting to note the difference between Susan's memory of how she got interested in chess and what we know about her parents' plans for her. László and Klara had already decided that Susan would become a top-ranked chess player, so they would have hardly counted on her just happening to find the chess pieces and becoming fascinated with them.

The precise details are not important, however. What is important is that Susan became interested in chess as a child — and that she became interested in the only way that a child of that age (she was three at the time) could become interested: she saw the chess pieces as fun. As toys. As something to play with. Young children are very curious and playful. Like puppies or kittens, they interact with the world mostly through play. This desire to play serves as a child's initial motivation to try out one thing or another, to see what is interesting and what is not, and to engage in various activities that will help them build their skills. At this point they're developing simple skills, of course — arranging chess pieces on a chessboard, throwing a ball, swinging a racket, organizing marbles by shape or pattern — but for future experts, this playful interaction with whatever has caught their interest is their first step toward what will eventually become their passion.

In the early 1980s the psychologist Benjamin Bloom directed a project at the University of Chicago that asked a simple question:

What does one find in the childhood of people who become experts that explains why they, among all people, develop such extraordinary abilities? The researchers working with Bloom chose 120 experts in six fields — concert pianists, Olympic swimmers, tennis champions, research mathematicians, research neurologists, and sculptors — and looked for common factors in their development. This study identified three stages that were common to all of them and that indeed appear to be common to the development of expert performers in every area, not just the six fields that Bloom and his colleagues examined.

In the first stage, children are introduced in a playful way to what will eventually become their field of interest. For Susan Polgár it was finding the chess pieces and liking their shapes. In the beginning, they were nothing more than toys to play with. Tiger Woods was given a little golf club to hold when he was just nine months old. Again, a toy.

In the beginning, a child's parents play with their child at the child's level, but gradually they turn the play toward the real purpose of the "toy." They explain the special moves of the chess pieces. They show how the golf club is used to hit the ball. They reveal the piano's ability to produce a tune rather than just a racket.

At this stage, the parents of children who are to become experts play a crucial role in the child's development. For one thing, the parents give their children a great deal of time, attention, and encouragement. For another, the parents tend to be very achievement-oriented and teach their children such values as self-discipline, hard work, responsibility, and spending one's time constructively. And once a child becomes interested in a particular field, he or she is expected to approach it with those same attributes — discipline, hard work, achievement.

This is a crucial period in a child's development. Many children will find some initial motivation to explore or to try something because of their natural curiosity or playfulness, and parents have an opportunity to use this initial interest as a springboard to an activity, but that initial curiosity-driven motivation needs to be supplemented. One excellent supplement, particularly with smaller children, is praise. Another motivation is the satisfaction of having developed a certain skill, particularly if that achievement is acknowledged by a parent. Once a child can consistently hit a ball with a bat, say, or play a simple tune on the piano or count the number of eggs in a carton, that achievement becomes a point of pride and serves as motivation for further achievements in that area.

Bloom and his colleagues found that often the experts in their study had picked up the particular interests of their parents. Parents who were involved in music, whether as performers or ardent listeners, often found their children developing an interest in music, as it was a way they could spend time with the parents and share the interest. Ditto for parents who immersed themselves in sports. The parents of children destined for more intellectual pursuits — such as the future mathematicians and future neurologists — were more likely to discuss intellectual topics with their kids, and they emphasized the importance of school and learning. In this way, the parents — at least the parents of children who would go on to be experts — shaped the interests of their children. Bloom reported no cases like the Polgárs, in which the parents set out consciously to push their children in a particular direction, but it doesn't have to be conscious. Simply by interacting strongly with their children, parents motivate their children to develop similar interests.

In this first stage, the children don't practice per se — that will come later — but many children do manage to come up with activities that are part play, part training. A good example is Mario Lemieux, widely considered to be one of the best hockey players to ever take to the ice. He had two older brothers, Alain and Richard, and the three of them would regularly go down to the basement of the family home, where they would slide around on their socked feet as if on ice skates and knock a bottle cap around with wooden kitchen spoons. Another is the British hurdler David Hemery — one of the best British track athletes ever — who turned many of his childhood activities into competitions with himself, challenging himself to constantly improve. When given a pogo stick for Christmas, for example, he stacked telephone books in order to practice jumping over obstacles. Although I don't know of any studies that look at the value of this sort of play practice, it seems likely that these children were taking their first steps down the path toward expertise.

Mario Lemieux's experience points up another salient feature of prodigies' early experience — how many of them had older siblings to be inspired by, to learn from, to compete with, and to model themselves after. Judit Polgár had Susan and Sofia. Wolfgang Mozart had Maria Anna, who was four and a half years older and who was already playing the harpsichord when Wolfgang first became interested in music. Tennis great Serena Williams followed in the footsteps of her sister Venus Williams, who was herself one of the best tennis players of the current era. Mikaela Shiffrin, who became the youngest slalom champion in history during the Olympic Games in 2014, had an older brother, Taylor, who was a competitive skier. And so on.

This is yet another sort of motivation. A child who sees an older sibling performing an activity and getting attention and praise from a parent will naturally want to join in and garner some attention and praise as well. For some children, competition with the sibling may itself be motivating, too.

In many of the cases that have been studied, children with talented siblings also had one or both parents encouraging them as well. The Polgár sisters we know about, and Mozart too: his father was not far behind László Polgár in his focus on developing a prodigy. Similarly, Serena and Venus Williams's father, Richard Williams, started them on tennis with the intention of turning them into tennis professionals. In such cases it can be hard to disentangle the influence of the siblings from that of the parents. But it is probably no coincidence in these cases that it is generally the younger siblings who have reached greater heights. Part of it may be that the parents learn from their experiences with the older siblings and do a better job with the younger ones, but it is also likely that the presence of an older sibling fully engaged in an activity provides a number of advantages for the younger sibling. By watching an older sibling engaging in an activity, a younger child may become interested in — and get started on — that activity much sooner than he or she might otherwise. The older sibling can teach the younger one, and it can seem more like fun than lessons provided by the parent. And competition between siblings will likely be more helpful to the younger sibling than the older one because the older one will naturally have greater skills, at least for a number of years.

Bloom found a slightly different pattern in the early days of the children who would grow up to be mathematicians and neurologists than in the athletes, musicians, and artists. In this case the parents didn't introduce the children to the particular subject matter but rather to the appeal of intellectual pursuits in general. They encouraged their children's curiosity, and reading was a major pastime, with the parents reading to the children early on, and the children reading books themselves later. They also encouraged their children to build models or science projects — activities that could be considered educational — as part of their play.

But whatever the specific details, the general pattern with these future experts was that at some point they became very interested in a particular area and showed more promise than other children of a similar age. With Susan Polgár that point came when she lost interest in the chess pieces simply as toys and became intrigued by the logic of how they moved around the board and interacted with other pieces during a game. At such a point, a child is ready to move to the next stage.

BECOMING SERIOUS

Once a future expert performer becomes interested and shows some promise in an area, the typical next step is to take lessons from a coach or teacher. At this point, most of these students encounter deliberate practice for the first time. Unlike their experiences up to this point, which have been mainly playful activities, their practice is about to become work.

In general, the instructors who introduce the students to this sort of practice are not experts themselves, but they are good at working with children. They know how to motivate their students and keep them moving forward as they adapt to the work of improving through deliberate practice. These teachers are enthusiastic and encouraging and reward their students — with praise or sometimes more concretely with candy or other small treats — when the students have accomplished something.

In the case of the Polgár sisters, László was their first teacher. He was not a particularly strong chess player — all of his daughters surpassed him well before their teenage years — but he knew enough to give them a good start on chess, and, most importantly, he kept them interested in the game. Judit has said that her father was the best motivator she had ever met. And this is perhaps the most important factor in the early days of an expert's development — maintaining that interest and motivation while the skills and habits are being built.

Parents have an important role to play, as well. (In the Polgárs' case, of course, László was both parent and teacher.) Parents help their children establish routines — say, practicing the piano for an hour each day — and they give them support and encouragement and praise them for improvements. They will, when necessary, push the children to prioritize their practice above other activities: practice first, play later. And if the children struggle too much to maintain their practice schedule, the parents may step in with more extreme measures. Some parents of Bloom's future experts had to resort to tactics like threatening to cut off piano lessons and sell the piano or to no longer take the child to swim practice. Obviously all of the future expert performers decided at this juncture that they wanted to keep going. Others might choose otherwise.

While there are various ways that parents and teachers can motivate children, the motivation must ultimately be something that comes from within the child, or else it won't endure. Parents of small children can motivate them with praise and rewards, among other things, but eventually that will not be enough. One way that parents and teachers can provide long-term motivation is to help the children find related activities that they enjoy. For example, if a child discovers that he or she loves playing a musical instrument in front of an audience, that may be enough to motivate the child to put in the necessary practice. Helping children develop mental representations can also increase motivation by increasing their ability to appreciate the skill they are learning. Representations of music help a child better enjoy listening to music performances and, in particular, to enjoy playing one's favorite pieces for oneself in the practice room. Representations of chess positions lead to a greater appreciation of the beauty of the game. Representations of a baseball game allow a child to understand and admire the strategy that underlies the play.

Bloom found a different pattern of interest and motivation among children who would become mathematicians, in large part because they started much later in their areas of interest. Parents don't usually hire special tutors to instruct their six-year-olds in mathematics. Instead, the future mathematicians first encountered serious mathematics courses — such as algebra, geometry, and calculus — in middle school and high school, and it was often the teachers in these courses, rather than their parents, who first stoked what would become their lifetime passion. The best teachers didn't focus on the rules for solving particular problems but rather encouraged their students to think about general patterns and processes — the *why* more than the *how*. This was motivating to these children because it sparked an intellectual interest that would drive their studies and, later, their research as mathematicians.

Because these children were older and had become interested enough in the subject independent of their parents' influence, they needed little parental goading or encouragement to do homework and whatever else the teacher might suggest. One thing their parents did do was to emphasize the importance of academic success in general and to make clear their expectations that their children would continue their schooling beyond high school and even beyond college.

During the first part of this stage, the encouragement and support of parents and teachers was crucial to the child's progress, but eventually the students began to experience some of the rewards of their hard work and became increasingly self-motivated. A piano student performed for others and appreciated the applause. A swimmer basked in the approval and respect of peers. These students became more vested in the process, and their self-image started to include those abilities that were setting them apart from their peers. In the case of team sports, like swimming, the students often relished being part of a group of like-minded people. But whatever the reasons, the motivation started to shift from external to internal in origin.

Finally, as the students continued to improve, they started to seek out better-qualified teachers and coaches who would take them to the next level. Piano students, for example, tended to move from a nearby teacher to the best teacher they could possibly reach, someone who often required an audition before accepting a student. Similarly, the swimmers would seek out the best coaches they could find, rather than the most conveniently located. With the step-up in the level of instruction, the students also began to practice longer hours. The parents still provided support, such as paying for lessons and equipment, but the responsibility for the practice shifted almost entirely to the students themselves and their coaches and teachers.

David Pariser, a researcher at Concordia University in Montreal, found a similar motivation in children who grew up to be gifted artists. They had a "self-fueling, self-motivating drive for tremendous work," he reported, although they still required "emotional and technical support" from their parents and teachers.

Bloom found that after two to five years at this stage, the future experts began to identify themselves more in terms of the skill they were developing and less in terms of other areas of interest, such as school or social life. They saw themselves as "pianists" or "swimmers" by the age of eleven or twelve or as "mathematicians" before they turned sixteen or seventeen. They were becoming serious about what they did.

Throughout these stages — and, indeed, throughout a person's life — it is difficult to untangle the various influences on motivation. There are certainly some intrinsic psychological factors that play a role, such as curiosity, and extrinsic factors, such as the support and encouragement of parents and peers. But too often we fail to acknowledge the neurological effects of actually doing the activity. We know that any sort of extended practice — playing chess or a musical instrument, learning mathematics, and so on — produces changes in the brain that lead to increased abilities in the skill being practiced, so it's reasonable to ask whether such practice also may produce changes in the brain structures that regulate motivation and enjoyment.

We can't answer that question yet, but we do know that people who develop skills in a certain area through years of practice seem to get a great deal of pleasure from engaging in that skill. Musicians enjoy performing music. Mathematicians enjoy doing mathematics. Soccer players enjoy playing soccer. Of course, it is possible that this is completely due to a self-selection process — that the only people who would spend years practicing something are those who naturally love to do it — but it is also possible that the practice itself may lead to physiological adaptations that produce more enjoyment and more motivation to do that particular activity. That is nothing but speculation at this point, but it is reasonable speculation.

COMMITMENT

Generally when they're in their early or mid teens, the future experts make a major commitment to becoming the best that they can be. This commitment is the third stage.

Now students will often seek out the best teachers or schools for their training, even if it requires moving across country. In most cases that teacher will be someone who has reached the highest levels in the field him- or herself—a concert pianist turned teacher, a swimming coach who has trained Olympic athletes, a top research mathematician, and so on. It is generally not easy to be accepted into these programs, and acceptance means that the teacher shares the student's belief that he or she can reach the highest levels.

The student faces expectations that gradually increase until the student is, in essence, doing as much as is humanly possible to improve. Swimmers are pushed to constantly improve on their personal best and, ultimately, to pursue national and even international record times. Pianists are expected to perfect their performance on increasingly difficult pieces. Mathematicians are expected to demonstrate their mastery of an area by working on a problem that no one has ever solved before. None of this is expected immediately, of course, but it is always the ultimate goal — to get out to the edge of human ability and rank among the best.

During this stage, the motivation lies solely with the student, but the family may still play an important support role. In the case of teenagers who move across country to train with a top coach, for instance, the family will often move, too. And the training itself can be incredibly expensive — not just the cost of the teacher or coach, but equipment, transportation, and so on.

In 2014, *Money* magazine estimated how much it cost a family to train a child who was an elite tennis player. Private lessons will cost $4,500 to $5,000 plus another $7,000 to $8,000 for group lessons. Court time will run you from $50 to $100 an hour. The entrance fee for a national tournament is about $150 plus transportation costs, and the best players compete in twenty or so tournaments a year. Bringing your coach along will cost another $300 a day plus transportation, lodging, and meals. Add all that up, and it's easy to spend $30,000 a year. But many of the really serious students head off to tennis academies where they train year-round, which can increase expenses dramatically. Attending the IMG Academy in Florida, for instance, will cost you $71,400 a year for tuition, room, and board—and you still have to pay to attend whatever tournaments you choose to play in.

Not surprisingly, Bloom reported that very few families could afford to have more than one child pursue this level of performance. Not only is it expensive, but it can be almost a full-time job for a parent to support the student in this pursuit—driving to and from practice during the week, providing transportation to competitions on the weekend, and so on.

However, the student who makes it to the end of this arduous road will have joined an elite cadre of people who can say categorically that they have reached the pinnacle of human achievement.

THE BENEFITS OF STARTING YOUNG

In Bloom's study, all 120 experts had begun their climb toward that pinnacle as children, which is typical among expert performers. But people frequently ask me what the possibilities are for someone who doesn't begin training until later in life. While the specific details vary by field, there are relatively few absolute limitations on what is possible for people who begin training as adults. Indeed, the practical limitations—such as the fact that few adults have four to five hours a day to devote to deliberate practice—are often more of an issue than any physical or mental limitations.

However, expertise in some fields is simply unattainable for anyone who doesn't start training as a child. Understanding such limitations can help you decide which areas you might wish to pursue.

The most obvious performance issues are those that involve physical abilities. In the general population physical performance peaks around age twenty. With increasing age we lose flexibility, we become more prone to injury, and we take longer to heal. We slow down. Athletes typically attain their peak performance sometime during their twenties. Professional athletes can remain competitive in their thirties or even early forties, with recent advances in training. In fact, people can train effectively well into their eighties. Much of the age-related deterioration in various skills happens because people decrease or stop their training; older people who continue to train regularly see their performance decrease much less. There are master's divisions in track and field competitions with age brackets up to eighty and beyond, and the people who train for these events do so in precisely the same way that people who are decades younger do; they just train for shorter periods with less intensity because of the increased risk of injury and the increased amount of time the body takes to recover from training. And with the realization that age is not the limitation it was once thought to be, more and more older adults are training harder and harder. Indeed, during the last few decades the performance of master athletes has improved at a much higher rate than that of younger athletes. Today, for example, a quarter of marathon runners in their sixties can be expected to outperform more than half of their competitors between the ages of twenty and fifty-four.

One of the oldest people to participate in these master's events is Don Pellmann, who in 2015 became the first person 100 years old or older to run one hundred meters in less than twenty-seven seconds. At the same track and field event — the San Diego Senior Olympics — Pellmann set four other age-group records as well — for the high jump, long jump, discus, and shot put. There are a number of athletes competing in Pellmann's current age group, which includes competitors from 100 to 104, and the competitions include most of the events in any track and field competition, including the marathon. (The world record time for the marathon in this age group is 8 hours, 25 minutes, 17 seconds, set by Fauja Singh of the United Kingdom in 2011.) The times may be longer, the distances jumped may be shorter, and the heights cleared may be lower, but these athletes are still going.

In addition to the gradual deterioration in physical abilities that accompanies aging, some physical skills simply cannot be developed to expert levels if one doesn't start working on them in childhood. The human body is growing and developing through adolescence up to the late teens or early twenties, but once we hit twenty or so, our skeletal structure is mostly set, which has implications for certain abilities.

For example, if ballet dancers are to develop the classic turnout — the ability to rotate the entire leg, beginning at the hip, so that it points directly to the side — they must start early. If they wait until after their hip and knee joints calcify — which typically happens between the ages of eight and twelve years — they'll probably never be able to get a full turnout. The same sort of thing is true for the shoulders of athletes, like baseball pitchers, whose sport requires them to throw a ball with an overhead motion. Only those who start training at an early age will have the requisite range of motion as adults, with the throwing arm able to be stretched well back behind the shoulder to produce the classic wind-up. And something similar holds true with the motion tennis players use when serving — only those who start young have the full range of the serving motion.

Professional tennis players who start young also overdevelop the forearm they use to hold the racket — not just the muscles, but the bones as well. The bones in a tennis player's dominant arm can be 20 percent thicker than the bones in his or her other arm, a huge difference that allows the bones in the dominant arm to endure the steady jolting that comes with hitting a tennis ball that may be traveling as fast as fifty miles per hour. However, even tennis players who start later in life — in their twenties — still can adapt to some degree, but just not as much as those who start younger. In other words, our bones retain their ability to change in response to stress well beyond puberty.

We witness this pattern again and again when we examine the relationship between age and the body's ability to adapt to stresses or other stimuli. The body and the brain both are more adaptable during childhood and adolescence than they are in adulthood, but in most ways they remain adaptable to some degree throughout life. The relationship between age and adaptability varies considerably according to exactly which characteristic you have in mind, and the patterns are very different for mental adaptations than for physical ones.

Consider the various ways that musical training can affect the brain. Studies have shown that some parts of the brain are larger in musicians than in nonmusicians, but there are certain parts of the brain for which this is true only if the musician began studying music as a young child. Researchers have found proof of this, for example, in the corpus callosum, the collection of tissue that connects the brain's hemispheres and serves as the communications path between them. The corpus callosum is significantly larger in adult musicians than in adult nonmusicians, but a closer look finds that it is really only larger in musicians who started practicing before they were seven. Since the initial publication of these findings in the 1990s, research has uncovered a number of other regions of the brain that are larger in musicians than in nonmusicians, but only if the musicians started training before a certain age. Many of those regions are related to muscle control, such as the sensorimotor cortices.

On the other hand, some parts of the brain involved in the control of movements, such as the cerebellum, are larger in musicians than in nonmusicians but show no difference in size between musicians who started music training later and those who started earlier. We don't know exactly what is going on in the cerebellum, but the implication would seem to be that musical training can affect the cerebellum in a noticeable way even when the training starts after childhood.

How adult brains learn is a relatively new and rather exciting field of study, and it is upending traditional beliefs that our brains become static once adolescence ends. The general lesson is that we can certainly acquire new skills as we age, but the specific way in which we acquire those skills changes as we get older. The human brain has the greatest amount of gray matter — the tissue containing neurons, the nerve fibers that connect the neurons, and the neurons' support cells — during early adolescence, at which point the brain starts par-ing down that gray matter. The synapses, the junctions between nerve cells, reach a maximum number early in life; a two-year-old child has about 50 percent more synapses than an adult. The specific details are not so important to us here as the general fact that the brain is constantly developing and changing through the first couple of decades of life, and so the background against which learning takes place is also changing. Thus it makes sense that a six-year-old's brain learns differently than a fourteen-year-old's brain, which learns differently than an adult brain — even if all of them are learning the same thing.

Consider what happens to the brain when it learns multiple languages. It is well known that people who speak two or more languages have more gray matter in certain parts of the brain — in particular, the inferior parietal cortex, which is known to play a role in language — and that the earlier a person learned a second language, the more extra gray matter there is. Thus, learning languages early in life takes place, it seems, at least in part through adding gray matter.

But a study of multilingual people who as adults studied to become simultaneous interpreters found a very different effect on the brain. These simultaneous interpreters actually had less gray matter than people who could speak the same number of languages but who didn't work as simultaneous interpreters. The researchers who carried out the study speculated that this disparity was because of the different contexts in which the learning took place. When children and adolescents learn new languages, it is against the backdrop of increasing gray matter, and so their learning the additional languages may occur through the addition of gray matter, but when adults continue their focus on multiple languages — this time with an emphasis on simultaneous translation — it is against a backdrop of pruning synapses. Thus the language learning that takes place in adulthood may take place more through getting rid of gray matter — getting rid of some inefficient nerve cells to speed up processes — which would explain why the simultaneous interpreters had less gray matter than other multilingual adults.

At this point, there are more questions than there are answers about the differences in learning among brains of various ages, but for our purposes there are two lessons to carry away: First, while the adult brain may not be as adaptable in certain ways as the brain of the child or adolescent, it is still more than capable of learning and changing. And second, since the adaptability of the adult brain is different from the adaptability of the young brain, learning as an adult is likely to take place through somewhat different mechanisms. But if we adults try hard enough, our brains will find a way.

MORE LESSONS FROM PERFECT PITCH

As an example of how the adult brain can find a way, consider perfect pitch — the example of brain adaptability that we began this book with. As I discussed, it seems that there is an age past which it is very difficult, if not impossible, to develop perfect pitch. If you do the proper training before you turn six, you are more likely to develop perfect pitch. If you wait until you're twelve, you'll be out of luck. At least that is the standard telling of the tale. It turns out that there is one more twist, and it is a very instructive one.

In 1969 Paul Brady, a researcher at the old Bell Telephone Laboratories, set out on what must have appeared to most to be a quixotic undertaking. At the time he was thirty-two years old, and he had been involved in music all his life. He'd played the piano since he was seven, he had sung in choruses since he was twelve, and he even tuned his own harpsichord. But he had never had perfect pitch or anything close. He had never been able to tell which note on the piano or harpsichord was being played. And because he was an adult, everything that was known about perfect pitch at the time indicated that he had lost his chance — he would never develop perfect pitch, no matter how hard he tried.

But Brady was not the sort to believe something was true just because everyone said it was. When he was twenty-one he decided he would try to teach himself to recognize notes. For two weeks he would play an A on his piano and try to remember what it sounded like. No luck. When he would come back a while later, he couldn't tell an A from a B or a C or a G-sharp. A few years later he tried again with a similar technique and similar results.

When he was thirty-two he decided to try again, this time vowing that he would keep working at it until he succeeded. He tried everything he could think of: He spent hours thinking about notes and playing pieces in his head, trying to hear what distinguished one note from another. Nothing. He tried playing piano pieces in different keys with the hope that he could learn to tell the difference between the different keys. Still nothing. After three months he was no closer to perfect pitch than when he had started.

Then he was inspired by a paper that described a training technique that had helped musicians without perfect pitch learn to recognize a single note. Brady set up a computer to produce random pure tones — these are tones that consist of a single frequency, unlike a note from a piano, which has a dominant frequency but also a number of other frequencies as well — and he used those pure tones to practice. At first, he had a large percentage of the randomly generated tones at the frequency of a C note, theorizing that if he could learn to recognize the C, he could use it as a base from which to recognize the other tones by their relationship to the C. Over time, as he got better and better at recognizing the C, the computer was set up to generate fewer and fewer of the Cs until all twelve notes were being generated with equal frequency.

Brady spent a half hour each day training with the tone generator, and at the end of two months he could identify every one of the twelve notes being played without error. Then, to test whether he had actually trained himself to have perfect pitch, he devised a test with a piano. Each day his wife would play one random note on the piano, and he would try to identify it. She did this for almost two months — fifty-seven days, to be exact—and at the end Brady looked to see how he had done. He had gotten thirty-seven exactly right; he'd missed eighteen by just half a tone—a B-flat instead of a B, for instance—and two by a full tone. Not perfect, but pretty close. Furthermore, the technical definition of perfect pitch actually allows a certain percentage of answers that are off by half a tone, and many people who are accepted by researchers as having perfect pitch actually make such errors themselves. So, by the literal definition of perfect pitch—and by any practical definition as well—Brady had taught himself, with two months of the right sort of practice, to have perfect pitch.

The article that Brady wrote describing his accomplishment got relatively little attention over the following decades, probably because he was just one person and he had done the experiment on himself, and researchers continued to assert that there was no convincing evidence that adults could develop perfect pitch.

In the mid 1980s a graduate student at Ohio State University, Mark Alan Rush, set out to test that claim with a carefully controlled study that attempted to develop perfect pitch in a group of adults. He decided to use a system designed by David Lucas Burge, who offered a training course that he claimed could help anyone develop perfect pitch. The course—which is still being sold today—spoke about the "colors" of different notes and asked students to listen to notes in such a way that they were paying attention not to such things as the loudness or timbre of the notes, but rather their color. Rush recruited fifty-two undergraduate music majors, half of whom would take Burge's course in an effort to develop perfect pitch, and half of whom would do nothing. Rush tested their ability to identify notes before and after a nine-month period, during which half of the students were working with Burge's course.

Rush's results were not exactly a ringing endorsement of Burge's methods, but they offered encouraging evidence about the possibility of improving one's ability to recognize notes. At the end of the nine-month period, the control group's scores were, not surprisingly, pretty much identical with their scores beforehand. But among the other group, a number of undergraduates had improved their judgment about notes. The test involved a total of 120 notes, and Rush kept track both of how many notes they got right and how far off they were on their wrong answers.

The student who saw the biggest improvement was also the student who started off with the best ear. That student got about 60 right on the first test and more than 100 right on the second — good enough to be described as having perfect pitch, but then that student had been well on his way before the training. Three other students who had relatively poor scores on the first test got much better on the second, doubling or tripling the number of right answers and making far fewer significant errors. The rest of the twenty-six students improved slightly or stayed the same. But it was clear from the pattern of improvement that the skill of recognizing notes could indeed be trained in adults — at least some adults — and that if the training were continued, or if perhaps a more effective approach were used, a number of those subjects might well have developed perfect pitch.

This is a very different picture than the traditional one, which sees perfect pitch as an either/or proposition: either you develop it as a child, or you never will. It may require a great deal of work, and it's possible that some adults may still never be able to do it, but it now appears that at least some adults can develop perfect pitch.

PATHBREAKERS

In 1997, a New Zealander named Nigel Richards entered his country's national Scrabble championship. To everyone's surprise, he won. Two years later he entered the World Scrabble Championship in Melbourne, Australia. He won again. Richards went on to dominate competitive Scrabble. He has won the world championship three times, the U.S. National Championship five times, the U.K. Open six times, and the King's Cup in Bangkok — the largest Scrabble competition in the world — twelve times. He has achieved the highest Scrabble rating ever. And perhaps most remarkable of all, he won the 2015 French Scrabble championship without even being able to speak the language. He took nine weeks to memorize the words from the French Scrabble dictionary, and he was ready.

The Scrabble world had never seen anything quite like Nigel Richards. But other fields certainly have. Many of the names are familiar — Beethoven, van Gogh, Newton, Einstein, Darwin, Michael Jordan, Tiger Woods. These are the people whose contributions leave their fields forever changed, the pathfinders who lead the way into new territory so that others can follow. This is the fourth stage of expert performance, where some people move beyond the existing knowledge in their field and make unique creative contributions. It is the least well understood of the four stages and the most intriguing.

One thing we do know about these innovators is that they, almost without exception, have worked to become expert performers in their fields before they started breaking new ground. It makes sense that this should be so: After all, how are you going to come up with a valuable new theory in science or a useful new technique on the violin if you are not intimately familiar with — and able to reproduce — the accomplishments of those who preceded you?

This is true even in those fields where it might not be so obvious that new inventions are always built upon older ones. Take Pablo Picasso. Someone who only knows his later, more famous paintings could reasonably conclude that they must have sprung directly from a mind untouched by earlier artistic traditions, because they looked so unlike anything from those traditions. In reality, Picasso began painting in an almost classical style — a style at which he was very accomplished. Over time he explored various other artistic styles, then combined them and modified them to develop his own style. But he had worked long and hard to develop himself as a painter and excel at the techniques his predecessors had mastered.

But where does such creativity ultimately come from? Is it not a whole other level beyond deliberate practice — which is, after all, based on practicing things in ways that other people have figured out in order to develop skills of the sort that others have already developed?

I don't believe so. Having studied many examples of creative genius, it's clear to me that much of what expert performers do to move the boundary of their fields and create new things is very similar to what they were doing to reach that boundary in the first place.

Consider this: Those experts who are at the very boundary of their professions — the best mathematicians, the top-ranked grandmasters in the world, the golfers who win major tournaments, the international touring violinists — didn't achieve their heights just by imitating their teachers. For one thing, most of them at this stage have already surpassed their teachers. The most important lesson they gleaned from their teachers is the ability to improve on their own. As part of their training, their teachers helped them develop mental representations that they could use to monitor their own performances, figure out what needs improving, and come up with ways to realize that improvement. These mental representations, which they are constantly sharpening and augmenting, are what guides them toward greatness.

You can picture the process as building a ladder step by step. You climb as high as you can and build one more step at the top of the ladder, climb up one more step, build another step, and so on. Once you get to the edge of your field, you may not know exactly where you're headed, but you know the general direction, and you have spent a good deal of your life building this ladder, so you have a good sense of what it takes to add on one more step.

Researchers who study how the creative geniuses in any field — science, art, music, sports, and so on — come up with their innovations have found that it is always a long, slow, iterative process. Sometimes these pathbreakers know what they want to do but don't know how to do it — like a painter trying to create a particular effect in the eye of the viewer — so they explore various approaches to find one that works. And sometimes they don't know exactly where they're going, but they recognize a problem that needs a solution or a situation that needs improving — like mathematicians trying to prove an intractable theorem — and again they try different things, guided by what has worked in the past. There are no big leaps, only developments that look like big leaps to people from the outside because they haven't seen all of the little steps that comprise them. Even the famous "aha" moments could not exist without a great deal of work to build an edifice that needs just one more piece to make it complete.

Furthermore, research on the most successful creative people in various fields, particularly science, finds that creativity goes hand in hand with the ability to work hard and maintain focus over long stretches of time — exactly the ingredients of deliberate practice that produced their expert abilities in the first place. For example, a study of Nobel Prize winners found that they had generally published scientific papers earlier than most of their peers and that they published significantly more papers throughout their careers than others in their discipline. In other words, they worked harder than everyone else.

Creativity will always retain a certain mystery because, by definition, it generates things that have not yet been seen or experienced. But we do know that the sort of focus and effort that give rise to expertise also characterize the work of those pioneers who move beyond where anyone has been before.

A psychologist who studied the Scrabble abilities of Nigel Richards called this "the Nigel effect." The appearance of Richards on the Scrabble scene and his amazing success in tournaments — he has won about 75 percent of all tournament games he has played, an incredibly high number for anyone who plays regularly against the world's best — showed other Scrabble players what could be achieved in their game. Until Richards came along, no one realized it was possible to be this good, and it forced other Scrabble players to look for ways to increase their own skill levels.

No one knows exactly how Richards became so good — he is notoriously uninterested in talking about his training techniques or strategies — but part of it is clearly that he knows more words than any of his competitors. Other Scrabble players are working to catch up, either by memorizing lots of words themselves or with some other approach that neutralizes his advantage. At this writing, Richards is still on top, but over time, his peers will inevitably devise techniques to match and even surpass him — and the field will have moved forward.

That's how it always is. The creative, the restless, and the driven are not content with the status quo, and they look for ways to move forward, to do things that others have not. And once a pathfinder shows how something can be done, others can learn the technique and follow. Even if the pathfinder doesn't share the particular technique, as is the case with Richards, simply knowing that something is possible drives others to figure it out.

Progress is made by those who are working on the frontiers of what is known and what is possible to do, not by those who haven't put in the effort needed to reach that frontier. In short, in most cases — and this is especially true in any well-developed area — we must rely on the experts to move us forward. Fortunately for all of us, that's what they do best.

8

But What About Natural Talent?

WHENEVER I WRITE OR SPEAK about deliberate practice and expertise I am invariably asked, But what about natural talent?

In my articles and my talks I always offer the same basic message that I have here: Expert performers develop their extraordinary abilities through years and years of dedicated practice, improving step by step in a long, laborious process. There are no shortcuts. Various sorts of practice can be effective, but the most effective of all is deliberate practice. Deliberate practice takes advantage of the natural adaptability of the human brain and body to create new abilities. Most of these abilities are created with the help of detailed mental representations, which allow us to analyze and respond to situations much more effectively than we could otherwise.

Fine, some people will reply, we understand all that. But, even so, aren't there some people who don't have to work as hard and can still be better than everyone else? And aren't there some people who are born without any talent for something — say, music or math or sports — so that no matter how hard they try, they'll never be any good at it?

It is one of the most enduring and deep-seated of all beliefs about human nature — that natural talent plays a major role in determining ability. This belief holds that some people are born with natural endowments that make it easier for them to become outstanding athletes or musicians or chess players or writers or mathematicians or whatever. While they may still need a certain amount of practice to develop their skills, they need far less than others who are not as talented, and they can ultimately reach much greater heights.

My studies of experts point to quite a different explanation of why some people ultimately develop greater abilities in an area than others, with deliberate practice playing the starring role. So let's separate myth from reality by exploring the intertwined roles of talent and training in the development of extraordinary abilities. As we'll see, innate characteristics play a much smaller — and much different — role than many people generally assume.

THE MAGIC OF PAGANINI

Niccolò Paganini was the greatest violinist of his era, but even for him the story that got told and retold over the years seemed impossible to believe. Depending on which version of the story one hears, the venue was a packed concert hall or an outdoor space where Paganini was serenading a lady at the request of her gentleman friend, but the basic details remain the same.

Paganini was nearing the end of an exquisite piece, with the audience — hundreds of concertgoers or perhaps just one very lucky lady — caught up in its beauty, unaware of anything else, when one of the violin's four strings snapped. Violin strings in those days — two centuries ago — were made from the intestines of sheep and more likely to give way than today's strings, and as Paganini had been approaching the climax of the composition, the poor string couldn't stand up to his forceful playing. The audience was stricken, sad to see the sudden end of the piece, when to their relief Paganini kept playing. The beauty of the piece was no less on three strings than it had been on four. Then a second string snapped, and again he didn't stop. This time the audience's relief was mixed with disbelief. How could he coax that beautiful melody out of just two strings? The dexterity and flexibility required of the fingers on his playing hand were more than the audience imagined was possible for any musician, yet the sound did not suffer. Paganini's playing on two strings was superior to what any other violinist could offer with four.

And then . . . you guessed it, a third string broke. Yet Paganini was undaunted. He finished the piece on the one remaining string, his fingers a blur and the audience amazed.

I heard this tale from my father when I was about ten years old, and it seemed to me that if indeed Paganini had been able to do what the story said, he must have been born with some inexplicable capacity that was very rare, perhaps even unique to him. Later in life, after I had been studying deliberate practice for some years, I still remembered my father's story, and I set out to track down the details in order to understand how such a feat might have been possible.

The first thing you discover when you read about Paganini is that he was truly a groundbreaking violinist. He developed a number of new techniques that allowed him to play the violin in unprecedented ways. And he was a showman — he liked to do things to impress the audience, things that no other violinist did. But the key to understanding my father's tale came from an old scientific report I found that repeated an old story told by Paganini himself. It went like this:

Some two hundred years ago Paganini found himself giving regular performances in Lucca, a town in Italy where Napoleon Bonaparte — then the emperor of France — spent a great deal of time with members of his family. One lady who was a regular attendee at Paganini's performances had caught his eye, and as their attraction for one another intensified, Paganini decided to write a composition for her that he would play at an upcoming concert. It was to be called

"Love Scene," and the notes were to reflect the conversation of two lovers. Paganini came up with the idea of removing the middle two strings of the violin and playing the composition only on the upper and lower strings, with the low G string representing the man's voice and the high E string the woman's. Paganini described the dialogue between them in this way: "Now the strings had to chide, and now to sigh; they had to whisper, to moan, to frisk, to rejoice and, at the end, to exult. And at the final reconciliation the newly united pair perform a pas de deux, that closes with a brilliant coda."

Paganini's performance of this composition was a great success, and after the concert he received an unusual request. A female member of Napoleon's family, whom Paganini referred to only as "the Princess," asked if he might write a piece to be performed on just one string. Apparently she was rather sensitive to sound, and compositions on all four strings sometimes proved too much for her nerves. Paganini agreed and named the resulting composition for the G string "Napoleon" because the emperor's birthday was near. The audience appreciated that song as well, and Paganini became intrigued with the challenge of writing and performing pieces on just the one string.

Of course, because he was a showman, as Paganini began to introduce one-string compositions into his repertoire, he would not simply announce them as such. He developed an act in which he would break one string after another by applying excessive force until he was down to the G string, where he would finish the song. He would write the songs with this in mind—with most of the song written to be performed on all four strings, then a section for three, a section for two, and a final section for just the G string. Because the audience had not heard the songs before—this was long before recorded music, of course—they had no idea what they were supposed to sound like. They only knew that they were heavenly—and that in the case of one song, Paganini had finished the composition while dealing with three broken strings.

Paganini's ability to write and play a beautiful tune on one string of the violin should not be taken lightly. He was a master of the violin, and this was an ability that no other violinist of his time possessed. But the performance was not the magical feat his listeners had believed it to be. It was the product of long, careful practice.

One of the major reasons that people believe in the power of innate talent is the apparent existence of natural prodigies — people who, like Paganini, seem to display skills unlike anyone else's or who exhibit expertise with little or no training. If such natural prodigies do indeed exist, then there must be at least some people who are born with innate abilities that allow them to do things other people cannot.

As it happens, I have made it a hobby to investigate the stories of such prodigies, and I can report with confidence that I have never found a convincing case for anyone developing extraordinary abilities without intense, extended practice. My basic approach to understanding prodigies is the same as it is for understanding any expert performer. I ask two simple questions: What is the exact nature of the ability? and, What sorts of training made it possible? In thirty years of looking, I have never found an ability that could not be explained by answering these two questions.

There are far too many reputed natural prodigies for me to address more than a fraction of them here, and that is not the purpose of this book. But let's look at a few cases just to provide a taste of how seemingly magical abilities can quickly become more believable when examined through the lens of deliberate practice.

MOZART AND HIS LEGEND

More than 250 years after his birth, Mozart remains the ultimate example of an inexplicable prodigy, the sort of person who was so accomplished at such a young age that there seems to be no way to explain it other than to assume he was born with something extra.

We know from the historical record that at a very young age Mozart was impressing audiences across Europe with his playing of the harpsichord, clavichord, and the violin. Beginning when Wolfgang was just six, his father took him and his sister on a multiyear tour across Europe. In Munich, Vienna, Prague, Mannheim, Paris, London, Zurich, and a number of other cities, the three Mozarts — Wolfgang; his father, Leopold; and his sister, Maria Anna — played exhibitions for the elites of the day. And of course little Wolfgang, his legs dangling from the bench and his hands barely able to reach the keyboard, was the main attraction. The Europeans had never seen anything like him.

So his abilities at a young age are inarguable. We must then ask, How did he practice? and Can it explain these abilities? Mozart certainly could play the violin and keyboard instruments with a facility that eighteenth-century Europeans were not used to in one so young, but today, when we're accustomed to seeing five- and six-year-olds trained in the Suzuki method playing beautifully on the violin and piano, his achievements seem much less wondrous. Indeed, there are YouTube videos of four-year-olds playing the violin and the piano with amazing facility — better than most adults. Yet we don't immediately assume that these children were born with some superior musical talent. We've seen enough of these "prodigies" now to know that they have developed their abilities through intense practice beginning at age two or earlier.

Mozart, of course, did not have the advantage of the Suzuki method, but he did have a father who was every bit as dedicated to raising a musical prodigy as any modern Suzuki parent. Furthermore, as I mentioned in the introduction, not only had Leopold Mozart written a book about teaching music to youngsters and tested out his ideas on Wolfgang's older sister, but Leopold was one of the very first music teachers to push the idea of starting children's lessons at a very young age. Wolfgang probably began his own training before the age of four. Given what we know now, we can explain how Mozart could have developed his abilities at such a young age without resorting to some sort of exceptional innate talent.

So that explains his precocity as a musician. But his talents as a child composer, another part of his legend, can't be dispatched by pointing to the mundane origins of modern violin prodigies. According to many biographies, he first began composing music when he was six, and he was eight when he wrote his first symphony. He wrote an oratorio and several keyboard concertos at eleven and an opera at twelve.

What was Mozart's talent here, really? What exactly did he do? Once we have answered that question, then we will try to figure out how he did it.

First, it's worth noting that music training today is quite different than what Wolfgang's father put him through. Today, Suzuki music teachers focus on one aspect of music — performance on a single instrument — while Leopold Mozart not only taught Wolfgang multiple instruments, he also worked with him on listening to and analyzing music and on writing music. So from early on, Leopold was pushing Wolfgang to develop his composing skills.

More to the point, though, the claims of Mozart composing at six and eight years old are almost certainly overstated. First, we know that the early compositions Wolfgang supposedly wrote are actually in Leopold's handwriting. Leopold claimed that he was just cleaning up young Wolfgang's work, but we have no way of knowing how much of a given composition was Wolfgang's work and how much was that of Leopold — who, remember, was a composer himself and, furthermore, a frustrated musician and composer who had never gotten all of the acclaim he wanted. There are plenty of parents of elementary school children today who get overly involved in their children's science-fair projects. It would not be at all surprising if something like that happened with young Wolfgang's compositions — particularly given the fact that Leopold had given up his own career by that time and had hitched his success to that of his son.

This seems even more likely, given what we know about the piano concertos that Wolfgang "composed" at eleven. Although these were considered original compositions for many years, musicologists eventually realized that they were all based on relatively unknown sonatas written by others. It now seems most likely that Leopold had assigned these to Wolfgang as compositional exercises to get him comfortable with the structure of the piano concerto and that there is relatively little in them that is original to Wolfgang. Furthermore, the evidence suggests that even on these reworkings of other people's compositions, Wolfgang had a great deal of help from his father. The first serious compositions that we can attribute to Wolfgang Mozart with certainty were written when he was fifteen or sixteen — after more than a decade of serious practice under his father's tutelage.

So we have no solid evidence that he did compose any significant music on his own before he was a teenager, and good reason to believe he did not. And when he did unequivocally begin to compose music that was original and sophisticated, he had been training to compose for a decade or so. In short, while there is no doubt that Mozart would become an extraordinary musician and composer, there is no evidence for — and plenty of evidence against — the claim that he was a prodigy whose accomplishments cannot be understood as the result of practice and must therefore be attributed to innate talent.

I have found the same thing with every child prodigy I have looked into. A more current example is Mario Lemieux, the Canadian hockey player generally recognized as one of the best of all time. There are various stories — many of them that can be traced back to Lemieux's mother — of how the young Mario took to the ice like a fish to water, skating from the very beginning as if he had been born to it and showing up older children who had been skating for years. These stories in turn have led some to claim that Lemieux is an example of a person who was clearly born with a superior natural talent.

However, a little digging into Lemieux's childhood reveals a situation very similar to that of the young Wolfgang Mozart. As I mentioned in chapter 7 Mario was the third son in a hockey-mad family, and he grew up with his two older brothers teaching him about hockey and skating almost from the time he could walk. The three would play wooden-spoon hockey together in the basement, sliding around on their stocking feet, and later their father built a rink in their front yard on which they could practice their hockey. So focused were Mario's parents on encouraging this hockey practice that they would even create stretches of "ice" in their home, where the boys could skate after it got too dark to skate outside. They did this by bringing piles of snow into the house, packing it down on the floors of the front hallway, dining room, and living room, and keeping the door open so that the house would stay cold. The brothers could then skate from room to icy room, giving a whole new meaning to the term *home ice.* In short, the evidence is that, like Mozart, Lemieux had a lot of practice before people began noticing what a "natural" talent he had.

THE MAGICAL HIGH JUMPER

Perhaps the most dramatic recent example of a supposed sports prodigy is the high jumper Donald Thomas. His story was told by David Epstein in the book *The Sports Gene,* and because it is so arresting, it has since been retold many times. Here are the basics.

Donald Thomas, originally from the Bahamas, was a student at Lindenwood University in Missouri and a member of the junior varsity basketball team. He was playing basketball with a friend who was a high jumper on the track team, and he showed off with some amazing dunks. Later, in the cafeteria, he and his friend were trading good-natured insults, and his friend said to him something to the effect of, "Sure, you can dunk, but I bet you couldn't clear six feet six in the high jump." (That would be a decent jump at the college level — particularly for athletes in the lower-division colleges such as Lindenwood — but the best college-level high jumpers regularly clear seven feet.) So Thomas took him up on the dare.

The two went over to the university's field house, where Thomas's friend set the high jump bar at six feet six. Thomas, wearing his basketball shorts and shoes, cleared it easily. His friend put it up to six feet eight. Thomas cleared it. Then his friend put the bar all the way up to seven feet. When Thomas cleared that as well, his friend grabbed him and took him to see the school's track coach, who agreed to have him join the track team and jump in a meet that was coming up in two days.

At that meet, still wearing basketball shoes instead of track shoes, Thomas won the competition with a jump of 2.22 meters, or about 7 feet 3.4 inches — which was a record at Eastern Illinois University, where the meet was held. Two months later Thomas competed for the Bahamas in Melbourne, Australia, in the British Commonwealth Games, where he placed fourth, with a jump of 2.23 meters. He would later transfer to Auburn University and compete for its track team, and just a year after his gift for high jumping was discovered, he placed first in the World Championships in Athletics in Osaka, Japan, with a jump of 2.35 meters, or nearly 7 feet 8.6 inches.

In his book, Epstein dramatized Thomas's accomplishments by comparing him with Stefan Holm of Sweden, who had been training rigorously on the high jump ever since he was a kid and had logged more than twenty thousand hours of practice. Yet at the 2007 World Championships in Athletics he was beaten by Thomas, who, Epstein estimated, had only a few hundred hours of training.

There is clearly a fascination with this type of story, where someone seems to have come out of nowhere to excel as some sort of naturally gifted performer. And these days, because "the ten-thousand-hour rule" has become so well known, the stories are often written as "proof" that this rule is wrong. Donald Thomas or someone else shows us that it is indeed possible to become the best in the world without practicing that much, if only you are born with the right genes.

I get it. People want to believe that there is magic in life, that not everything has to abide by the staid, boring rules of the real world. And what could be more magical than being born with some incredible ability that doesn't require hard work or discipline to develop? There is an entire comic-book industry built on that premise — that sometimes something magical happens, and you suddenly acquire incredible powers. Unbeknownst to you, you were actually born on the planet Krypton and you can fly. Or you were bitten by a radioactive spider and you can cling to walls. Or you were exposed to cosmic radiation and now you can become invisible.

But my decades of research in the area of expertise have convinced me that there is no magic. By examining the case of someone with exceptional abilities through the lens of those two earlier questions I posed — What is the talent? What practice led to the talent? — you can pull back the curtain and find what is really going on.

Consider Thomas's story. There is actually little or nothing about his background that we know other than what has come from him directly, which has been very limited, so it is difficult to track down exactly what sorts of training he might have had. But we do know a few things. First, Thomas himself told an interviewer that he had competed in the high jump in at least one intramural meet in high school and had jumped "something like 6-2, 6-4, nothing memorable." So we know that he had at least competed in the high jump before, and if he was competing on his high school team, he almost certainly got some training. And Thomas is being a bit modest when he says the jump was "nothing memorable." While six feet four is by no means a great jump in high school, it is a good one.

Of course, it might be possible that Thomas had no training whatsoever in high school and simply went out one time and jumped six feet four without practice, just like he jumped seven feet without practice in college. The problem with this scenario is that we actually have photos of Thomas clearing the bar in the first college meet, and it is not the technique of someone who has never trained in the high jump. Thomas is clearly using "the Fosbury Flop," named for the American high jumper Dick Fosbury, who popularized it in the 1960s. The flop is a highly counterintuitive way to get over the bar: you run at the bar in a curving path so that as you reach the point directly in front of the bar, your back is facing it, and then you jump up and arch backward over the bar, throwing your feet up at the last minute so as to not knock the bar down. It is not enough to simply have a lot of spring in your legs; you have to use just the right technique to execute this jump. No one performs the Fosbury Flop effectively without extended practice. So although we don't know anything explicitly about Thomas's training before that day in the Lindenwood field house, we can be sure that he spent quite a few hours learning that technique to the point that he could clear "something like 6-2, 6-4."

The second thing we know is that Thomas had an incredible jumping ability when dunking. There are videos of him dunking a basketball after taking off from the free-throw line, fifteen feet from the goal, and flying over a couple of people on his way to the rim. Again, while we have no information on how much practice Thomas put into his dunking, we can be certain he worked hard to develop that spring in his legs. Dunking was clearly something that he was proud of, so it would be strange if he had not worked hard on it. So again, it is circumstantial, but it seems clear that Thomas practiced diligently on his ability to jump high on his dunks. And, as it happens, the sort of jumping technique you use in dunking — which involves taking several steps and then jumping off of one foot — is very similar to what is used in the high jump. By training his dunking ability, Thomas was also training for the high jump. A 2011 study shows that the ability to jump off of one leg is closely correlated with the height of a high jump among skilled high jumpers.

Third, it's worth noting that Thomas is six feet two, which is a good, if not ideal, height for high jumping. As I mentioned earlier, the only two areas where we know for certain that genetics affects sports performance are height and body size. Stefan Holm, the Swedish high jumper whom Thomas defeated in the 2007 World Championships, is only five feet eleven, extremely short for a high jumper. Holm had to train extra hard to make up for this deficit. Thomas was genetically gifted with a good body size for the high jump.

So when you put all this together, Thomas's feat no longer seems quite so magical — impressive, yes, but not magical. Thomas had almost certainly trained in the high jump previously, at least enough to develop a good Fosbury Flop, and he had developed his ability to jump high off of one foot through his practice dunking — an unusual approach to training for the high jump, but in Thomas's case at least, an effective one.

And we have one more bit of evidence. As of 2015 Thomas had been competing for nine years in the high jump. He has been training under coaches who know how to get the most out of an athlete. If he had indeed been nothing but raw potential in 2006, we should have seen some phenomenal growth from him since he started training rigorously. Indeed, in the first year or so after he was discovered, people were predicting that his innate talent meant that he must surely develop to the point that he would break the world record, which is 2.45 meters, or 8.04 feet. But he hasn't come close. His best jump in competition came in that 2007 World Championships in Athletics, when he cleared 2.35 meters. He has approached that height a few times since, but has never equaled it. In the 2014 Commonwealth Games he jumped 2.21 meters, less than he was able to jump eight years earlier in the 2006 Commonwealth Games, when he first made a name for himself. The most obvious conclusion to draw from this is that when Thomas first competed in college in 2006, he had already had a great deal of training — both high-jump training and training to jump higher for dunking — so it was difficult for further training to make a big difference. If he had indeed never trained, there should have been much more improvement.

SAVANTS

Besides apparent prodigies like Mozart or Donald Thomas, there is one other group of people who are often claimed to exhibit extraordinary abilities that seem to have appeared almost as if by magic, and these are people with savant syndrome. The abilities of these savants, as they are now called, generally arise in very specific areas. Some play a musical instrument and often have thousands of different pieces of music memorized and can sometimes play a new piece of music after hearing it once. Others can paint or sculpt or do other sorts of art, often producing incredibly detailed works. Some do arithmetic calculations, such as multiplying two large numbers in their heads. Still others do calendar calculations, such as stating what day of the week October 12, 2577, will be (a Sunday). What makes these abilities particularly noteworthy is that most of these savants are otherwise mentally challenged in one way or another. Some perform extremely poorly on IQ tests, while others are severely autistic and can barely interact with other people. The appearance of these striking abilities in people who otherwise struggle to function in the world is what makes the savant syndrome so intriguing — and also what makes it seem as if these abilities must have appeared without the normal sort of practice that we generally expect.

Again, the best approach to take in understanding these abilities is to first understand exactly what they are and then to look for the sorts of practice that could explain them. Research that has taken that approach indicates that savants are not the recipients of some miraculous talent; instead they have worked for it, just like anyone else.

Francesca Happé and Pedro Vital, two researchers at King's College London, compared autistic children who develop savantlike abilities with autistic children who did not develop such abilities. They found that the autistic savants are much more likely than the nonsavants to be very detail-oriented and prone to repetitive behaviors. When something captures their attention, they will focus on it to the exclusion of everything else around them, retreating into their own worlds. These particular autistic people are more likely to practice obsessively a musical piece or memorize a collection of phone numbers — and thus are likely to develop skills in those areas in the same way the people engaging in purposeful or deliberate practice do.

One of the best examples of this is Donny, an autistic savant who is the fastest, most accurate calendar calculator who has ever been tested. Donny can provide the day of the week for a particular date within a second of hearing the date, and he is almost invariably correct. Marc Thioux of the University of Groningen in the Netherlands has been studying Donny for a number of years, and Thioux's research gives us an unprecedented window into the mind of an autistic savant.

Donny is addicted to dates, Thioux has said. The first thing that Donny does when he meets someone is to ask for the person's birthday. He is constantly thinking about dates and repeating dates to himself. He has memorized all fourteen possible yearly calendars — that is, the seven normal-year calendars in which January 1 is a Sunday, Monday, Tuesday, Wednesday, Thursday, Friday, or Saturday, and the corresponding leap-year calendars — and he has developed ways to quickly calculate which of those fourteen possible calendars applies to any given year. When asked which day of the week a particular date will fall on, Donny focuses first on the year in order to figure out which of the fourteen calendars to use, and he then refers to that mental calendar to determine the day of the week for the date in question. In short, Donny possesses a highly developed skill that is the result of years of obsessive study, but no sign of a miraculous innate talent.

In the late 1960s, a psychologist named Barnett Addis set out to see if he could train someone of normal intelligence to do the same sorts of calendar calculations that savants do. In particular, he had been studying how two calendar-calculating twins performed their feats. The twins, who each had an IQ in the 60–70 range, were able to provide the days of the week for dates out to the year A.D. 132470 within an average of six seconds. Addis found that the twins' method seemed to involve finding an equivalent year between 1600 and 2000 and then adding up numbers that corresponded to the day of the month, the month, the year, and the century. With this understanding, Addis then trained a graduate student in that method to see if it actually worked. In just sixteen practice sessions the graduate student was able to calculate just as fast as either of the twins. Most interestingly, the graduate student took different amounts of time to generate the weekday, depending on the amount of calculation required. His pattern of response times matched those of the best twin, suggesting to Addis that the two of them were indeed getting their answers via similar cognitive processes.

The lesson here is that there is clearly nothing magical about Donny's — or any other savant's — calendar-calculation abilities. Donny developed his abilities over years of working with and thinking about dates, reaching the point where he knows each of the fourteen different calendars as well as you or I know our phone numbers, and he has developed his own technique — which, in this case, researchers still have not completely understood — for determining which calendar to use for which year. It is nothing that a motivated college student in a psychology experiment could not do.

We do not yet know exactly how other savants do what they do and how they have developed their particular skills — savants are generally hard to communicate with or to question about their methods — but, as I noted in a 1988 review, studies of savants' abilities indicate that these are primarily acquired skills, which in turn implies that the savants develop those abilities in ways that are very similar to how other experts do it. That is, they practice in a way that engages their brain's adaptability, which in turn changes their brains in ways that lead to their extraordinary abilities. More recent case studies of savants' brains have been consistent with this idea.

THE ANTI-PRODIGIES

I could keep going with more analyses of prodigies and savants, but it would just be more of the same. The bottom line is that every time you look closely into such a case you find that the extraordinary abilities are the product of much practice and training. Prodigies and savants don't give us any reason to believe that some people are born with natural abilities in one field or another.

But what about the flip side of prodigies? What about people who seem to have been born without any talent whatsoever in this field or that? On an individual level this is a very difficult issue to address since it can be hard to figure out exactly why a particular person didn't accomplish something. Was it for lack of effort, lack of adequate teaching, or lack of "innate talent"? You can't always tell, but consider the following cases.

About one-sixth of all American adults believe they can't sing. They can't carry a tune in a bucket. They couldn't hit a note if you gave them a tennis racket. And, generally speaking, these people aren't very happy about it. If you talk to music teachers or the few researchers who study nonsingers, they will tell you that these musically challenged sorts would like things to be different. At the very least they'd like to sing "Happy Birthday" without frightening people. They may even daydream about karaoke and bringing the house down with their versions of "My Way" or "Baby One More Time."

But somewhere along the way someone convinced them that they couldn't sing. Interviews have found that it was usually some sort of authority figure — a parent, an older sibling, a music teacher, maybe a peer they admired — and it usually came at some defining — and often painful — moment that they still remember quite well as adults. Most often they were told they were "tone-deaf." And so, believing they just weren't born to sing, they gave up.

Now the term *tone-deaf* actually has a very specific meaning: it means you can't tell the difference between one musical note and another. For example, if someone hits a C note on the piano and then a D note, a tone-deaf person can't tell the difference. And, of course, if you can't tell one note from the next, it would certainly be impossible to carry a tune, which is just a series of notes strung together. It would be like trying to paint a sunset when you can't tell red from yellow from blue.

Some people are indeed born tone-deaf. The medical condition is known as "congenital amusia," but here is the twist: it is exceedingly rare. It is so rare that the discovery of a woman with this condition rated an article in a major scientific journal. She had no obvious brain damage or defects, had normal hearing and intelligence, and yet she could not tell the difference between a simple melody she had already heard and a new one she had never heard before. Interestingly enough, she also had trouble distinguishing different musical rhythms. This woman, no matter how hard she tried, would never be able to carry a tune.

But that is not the case for most people. The major obstacle that people who believe they can't sing must overcome is that belief itself. Various researchers have studied this issue, and there is no evidence that large numbers of people are born without the innate ability to sing. Indeed, there are some cultures, such as the Anang Ibibio of Nigeria, where everyone is expected to sing, everyone is taught to sing, and everyone can sing. In our culture, the reason that most nonsingers cannot sing is simply that they never practiced in a way that led them to develop the ability to sing.

Could the same thing be true about a subject like math? There is perhaps no area in which more people will tell you, "I am no good in . . ." A large percentage of students, particularly in the United States, leave high school with the conviction that they just do not have the genetic endowment to do any math more complicated than addition, subtraction, and perhaps multiplication. But a number of successful efforts have shown that pretty much any child can learn math if it is taught in the right way.

Perhaps the most intriguing of these efforts is a curriculum called Jump Math, developed by John Mighton, a Canadian mathematician. The program uses the same basic principles found in deliberate practice: breaking learning down into a series of well-specified skills, designing exercises to teach each of those skills in the correct order, and using feedback to monitor progress. According to teachers who have used the curriculum, this approach has allowed them to teach the relevant math skills to essentially every student, with no one left behind. Jump was evaluated in a randomized controlled trial in Ontario with twenty-nine teachers and approximately three hundred fifth-grade students, and after five months the students in the Jump classes showed more than twice as much progress as the others in understanding mathematical concepts as measured by standardized tests.

Unfortunately, the results of the trial have not appeared in a peer-reviewed scientific journal, so it is hard to judge them objectively, and we will need to see the results reproduced in other school districts before we can trust them completely, but the results agree with what I have observed generally in a variety of fields, not just singing and math, but writing, drawing, tennis, golf, gardening, and a variety of games, such as Scrabble and crossword solving: People do not stop learning and improving because they have reached some innate limits on their performance; they stop learning and improving because, for whatever reasons, they stopped practicing—or never started. There is no evidence that any otherwise normal people are born without the innate talent to sing or do math or perform any other skill.

PRACTICE VERSUS "TALENT" IN CHESS

Think back to when you were a kid and you were just starting to learn to play the piano or to throw a baseball or to draw something. Or maybe think about how it felt when you were just a little further along—you'd been playing soccer for six months and it was just starting to make sense, or you had joined the chess club a year earlier and you had finally gotten a basic command of the game, or you had figured out addition and subtraction and multiplication and then your teacher was throwing long division at you. In all of these cases, when you looked around you would have noticed that some of your friends or classmates or peers were doing better than others, and some were doing worse. There are always obvious differences in how quickly different people pick something up. Some just seem to have an easier time playing a musical instrument. Some just seem to be natural athletes. Some just seem to be naturally good with numbers. And so on.

And because we see such differences in beginners, it's natural to assume that those differences will persist — that the same people who did so well in the beginning will continue to breeze through later on. These lucky people, we imagine, were born with innate talents that smooth the way and lead them to excel. This is an understandable result of observing the beginning of the journey and concluding that the rest of the journey will be similar.

It is also wrong. Once we take a look at the entire journey — from beginner to expert — we develop a very different understanding of how people learn and improve and what it takes to excel.

Perhaps the best example we have of this comes from chess. In the popular imagination, great skill in chess is intimately tied to tremendous logic and intellect. If an author or screenwriter wishes to signal that a character is particularly brilliant, that character will be seated at a chessboard and will checkmate his or her opponent with the proper savoir faire. Even better, this genius will come across a game already in progress and, after glancing at the board for a second or two, point out the winning line of play. Quite often the chess player is a quirky but brilliant detective, or perhaps the equally quirky and almost equally brilliant criminal mastermind — or preferably both, so that the opponents can face each other across the board, matching wits and trading witticisms. Sometimes, as in the climactic scene in the 2011 movie *A Game of Shadows* with Sherlock Holmes and Professor Moriarty, the two of them end up ignoring the chessboard altogether and just spit out their moves at each other like two boxers feinting and jabbing until one lands the knockout punch. But no matter the circumstances, the message is always the same: a mastery of chess signals the sort of deep intelligence that only a few are fortunate enough to be born with. And, conversely, playing chess brilliantly demands a brilliant mind.

And if you examine chess-playing ability in children who are just learning to play, those with higher IQs do indeed become better players faster. But that is just the beginning of the story — and it is the end of the story that truly tells the tale.

Over the years many researchers have examined the connection between intelligence and chess-playing ability. Some of the earliest work was done in the 1890s by Alfred Binet, the father of intelligence testing, who studied chess players mainly in an attempt to understand what sort of memory was required to play blindfold chess. Binet developed his IQ test as a method of identifying students who had problems doing well in school, and, indeed, he succeeded, as IQ tests are very much correlated with academic success. But since Binet's time many researchers have argued that the IQ test measures general abilities that are correlated with success in virtually any domain, such as music and chess. These researchers thus believe that IQ tests measure some sort of general innate intelligence. Others disagree, however, and argue that IQ is best thought of not as innate intelligence but rather simply as what IQ tests measure, which can include such things as knowledge about relatively rare words and acquired skills in mathematics. Without delving deeply into that debate, I will just say that I think it is best to not equate IQ with innate intelligence but simply to stick with the facts and think of IQ as some cognitive factor, measured by IQ tests, that has been shown to predict certain things, such as success in school.

Since the 1970s a growing number of researchers have been following in Binet's footsteps and trying to understand how chess players think and what makes a good chess player. One of the most enlightening of these studies was carried out in 2006 by three British researchers, Merim Bilalić and Peter McLeod of Oxford University and Fernand Gobet of Brunel University. For reasons that we will get to in a moment, the three chose to study not grandmasters but rather a collection of chess-playing schoolchildren, recruiting fifty-seven kids from chess clubs in primary and secondary schools. The young chess players were generally between nine and thirteen years old, and they had been playing the game for an average of about four years. Some of them were very good — good enough to easily beat the average adult who plays in chess tournaments — and some were not very good at all. Forty-four of the fifty-seven were boys.

The goal of the study was to examine what role — if any — IQ plays in how good a chess player someone can become. This is a question that quite a few psychologists had already examined, and, as the three researchers noted in the paper that they published reporting their results, the issue had been rather unsettled. For example, some research had found a relationship between IQ and chess-playing ability as well as between tests measuring visuospatial abilities and chess skill. Neither would seem particularly surprising, given the general view that chess requires higher-than-normal intelligence and given that visuospatial abilities would seem particularly important to chess, since chess players must be able to visualize chess positions and the movements of pieces as they examine potential lines of play. But these studies were done in young chess players, and while they found that these young players did have higher-than-average IQ scores, there was no clear relationship between IQ and how good a particular player was.

By contrast, studies done in adults have generally found adult chess players to have no better visuospatial abilities than normal non-chess-playing adults. Research has also shown that skilled adult chess players — even grandmasters — do not have systematically higher IQs than other adults with similar levels of education. Nor is there any correlation between the IQs of highly skilled chess players and their chess ratings. As strange as it seems to those of us who have grown up with the tortured-but-brilliant fictional characters who excel at chess, all of the evidence says that higher intelligence is not correlated with better chess playing among adults.

Even stranger is the case of Go, which has often been referred to as the Asian version of chess. It is played by two people who alternately place their stones — white for one player, black for the other — on one of the intersecting points on the 19 x 19 grid that makes up the board. The goal is to surround and capture the other player's stones, and the winner is the one who controls the larger area of the board at the end of the game. While there is only one type of piece and only one type of move — putting a stone on an intersection point — the game is actually more complex than chess, in the sense that there are far more different possible games that can be played, and, indeed, it has proved far more challenging than chess to develop software to play the game well. Unlike the best chess-playing computer programs, which can consistently beat chess grandmasters, the best Go programs — at least as this is written in 2015 — cannot stand up to top-ranked Go players.

Thus, as with chess, you might assume that Go masters must have high IQs or perhaps exceptional visuospatial skills, but again you would be wrong. Recent studies of Go masters have found that their average IQ is, if anything, below average. Two separate studies of Korean Go experts found an average IQ of about 93, compared with control groups of non-Go-playing Koreans matched for age and sex, which had an average IQ around 100. While the numbers of Go masters in the two studies were small enough that the below-average IQs could have been just statistical flukes, it is clear that Go masters, on average, score no higher on IQ tests than people in the general population.

Against this background the three British researchers set out to resolve the conflicting results on chess players. Does a higher intelligence (that is, a higher IQ score) help one develop a better chess game or not? The researchers' plan was to do a study that took into account both intelligence and practice time. Earlier studies had looked at one or the other but not both together.

Bilalić and his colleagues set out to learn as much as they could about their group of fifty-seven young chess players. They measured various aspects of their intelligence — not just their IQ and their spatial intelligence, but their memory, verbal intelligence, and speed of processing. They asked the children about when they'd started to play and how many hours they spent practicing. They also asked the kids to keep practice diaries for about six months, in which they recorded the amount of time they spent practicing each day. One weakness of the study is that much of the "practice" time was actually spent playing chess games against other members of their chess clubs rather than in solitary practice, and the researchers did not distinguish between the two types of practice. Still, the measures offered a reasonable estimate of how much effort each child had put into developing his or her game. Finally, the researchers assessed the kids' chess skills by giving them chess problems to solve and by briefly showing them chessboards with the pieces arranged in a position from the middle of a game and asking them to reconstruct the boards from memory. A few of the subjects were regularly participating in tournaments, and in these cases the researchers also had their chess ratings to work with.

When the researchers analyzed all their data, they found results similar to those seen by other researchers. The amount of chess practice that the children had done was the biggest factor in explaining how well they played chess, with more practice being correlated with better scores on the various measures of chess skill. A smaller but still significant factor was intelligence, with higher IQ being related to better chess skills. Surprisingly, visuospatial intelligence wasn't the most important factor, but rather memory and processing speed were. Looking at all their evidence, the researchers concluded that in children of this age, practice is the key factor in success, although innate intelligence (or IQ) still plays a role.

The picture changed dramatically, however, when the researchers looked at only the "elite" players in the group. These were twenty-three children — all boys — who were regularly playing in tournaments on the local, national, and sometimes international levels. They had an average chess rating of 1603, with the highest being 1835 and the lowest 1390. In short, these kids were already quite good at chess. The average chess rating for everyone who plays in chess tournaments, both adults and children, is about 1500, which means that most of the boys in the elite group were above that average, and even the worst of them would have had little trouble checkmating a competent adult player.

Among these twenty-three elite players the amount of practice was still the major factor determining their chess skills, but intelligence played no noticeable role. While the elite group did have a somewhat higher average IQ than the average IQ for the entire group of fifty-seven, the players in the elite group with lower IQs were, on average, slightly better players than those in the elite group with higher IQs.

Stop and digest that for a moment: among these young, elite chess players, not only was a higher IQ no advantage, but it seemed to put them at a slight disadvantage. The reason, the researchers found, was that the elite players with lower IQs tended to practice more, which improved their chess game to the point that they played better than the high-IQ elite players.

This study goes a long way toward explaining the apparent contradiction between the earlier studies, which had found that IQ was linked with greater chess skill in young players but not in adult tournament players and not in masters and grandmasters. And this explanation is very important to us because it applies not just to chess players but to the development of any skill.

When children are just beginning to learn chess, their intelligence — that is, their performance on IQ tests — plays a role in how quickly they can learn the game and reach a certain minimal level of competence. Kids with higher IQ scores generally find it easier to learn and remember rules and to develop and carry out strategies; all of these things give them an advantage in the early stages of learning the game, when one plays by abstract thinking applied directly to the pieces on the board. This type of learning is not all that different from the learning that goes on in schools, which was the target of Binet's original project developing IQ tests.

But we know that as children (or adults) study and learn the game, they develop sets of mental representations — in essence, mental shortcuts — that give them both a superior memory for the sorts of chess positions found in a game and an ability to quickly zero in on appropriate moves in a given situation. It seems quite likely that these superior mental representations allow them to play the game more quickly and powerfully. Now when they see a certain arrangement of pieces, they don't have to carefully figure out which piece is attacking or could attack every other piece; instead they recognize a pattern and know almost reflexively what the most powerful moves and countermoves would likely be. No longer do they have to apply their short-term memory and analytical skills to imagine what would happen if they made this move and their opponent made that move and so on, trying to recall the position of every piece on the board. Instead, they have a good general idea of what is going on in a given position — in terms of lines of force or whatever imaging technique they use — and they use their logical abilities to work with their mental representations, rather than the individual pieces on the board.

With enough solitary practice, the mental representations become so useful and powerful in playing the game that the major thing separating two players is not their intelligence — their visuospatial abilities, or even their memory or processing speed — but rather the quality and quantity of their mental representations and how effectively they use them. Because these mental representations are developed specifically for the purpose of analyzing chess positions and coming up with the best moves — remember, they are usually developed through thousands of hours of studying the games of grandmasters — they're far more effective for playing chess than simply using one's memory and logic and analyzing the collection of pieces on the board as individually interacting items. Thus, by the time one becomes a grandmaster or even an accomplished twelve-year-old tournament player, the abilities measured by IQ tests are far less important than the mental representations one has developed through practice. This explains, I believe, why we see no relationship between IQ and chess ability when we look at accomplished players.

Of course, the abilities measured by IQ tests do seem to play a role early on, and it seems that children with higher IQs will play chess more capably in the beginning. But what Bilalić and his colleagues found was that among the children who played in chess tournaments — that is, the chess players who were devoted enough to the game to take it to a level beyond playing in their school chess club — there was a tendency for the ones with lower IQs to have engaged in more practice. We don't know why, but we can speculate: All of these elite players were committed to chess, and in the beginning the ones with higher IQs had a somewhat easier time developing their ability. The others, in an effort to keep up, practiced more, and having developed the habit of practicing more, they actually went on to become better players than the ones with higher IQs, who initially didn't feel the same pressure to keep up. And here we find our major takeaway message: In the long run it is the ones who practice more who prevail, not the ones who had some initial advantage in intelligence or some other talent.

THE REAL ROLE OF INNATE CHARACTERISTICS

The results from the chess study provide a crucial insight into the interplay between "talent" and practice in the development of various skills. While people with certain innate characteristics — IQ, in the case of the chess study — may have an advantage when first learning a skill, that advantage gets smaller over time, and eventually the amount and the quality of practice take on a much larger role in determining how skilled a person becomes.

Researchers have seen evidence of this pattern in many different fields. In music, as in chess, there is an early correlation between IQ and performance. For example, a study of ninety-one fifth-grade students who were given piano instruction for six months found that, on average, the students with higher IQs performed better at the end of those six months than those with lower IQs. However, the measured correlation between IQ and music performance gets smaller as the years of music study increase, and tests have found no relationship between IQ and music performance among music majors in college or among professional musicians.

In a study on expertise in oral surgery, the performance of dental students was found to be related to their performance on tests of visuospatial ability, and the students who scored higher on those tests also performed better on surgical simulations done on the model of a jaw. However, when the same test was done on dental residents and dental surgeons, no such correlation was found. Thus the initial influence of visuospatial ability on surgical performance disappears over time as the dental students practice their skills, and by the time they have become residents, the differences in "talent" — in this case, visuospatial ability — no longer have a noticeable effect.

Among the people studying to be London taxi drivers that we discussed in chapter 2, there was no difference in IQ between the ones who finished the course and became certified drivers and those who dropped out. IQ made no difference in how well the drivers could learn to find their way around London.

The average IQ of scientists is certainly higher than the average IQ of the general population, but among scientists there is no correlation between IQ and scientific productivity. Indeed, a number of Nobel Prize–winning scientists have had IQs that would not even qualify them for Mensa, an organization whose members must have a measured IQ of at least 132, a number that puts you in the upper 2 percentile of the population. Richard Feynman, one of the most brilliant physicists of the twentieth century, had an IQ of 126; James Watson, the co-discoverer of the structure of DNA, had an IQ of 124; and William Shockley, who received the Nobel Prize in Physics for his role in the invention of the transistor, had an IQ of 125. Although the abilities measured by IQ tests clearly help performance in the science classroom, and students with higher IQs generally perform better in science classes than those with lower IQs — again consistent with Binet's efforts to measure school learning — among those who have become professional scientists, a higher IQ doesn't seem to offer an advantage.

A number of researchers have suggested that there are, in general, minimum requirements for performing capably in various areas. For instance, it has been suggested that scientists in at least some fields need an IQ score of around 110 to 120 to be successful, but that a higher score doesn't confer any additional benefit. However, it is not clear whether that IQ score of 110 is necessary to actually perform the duties of a scientist or simply to get to the point where you can be hired as a scientist. In many scientific fields you need to hold a Ph.D. to be able to get research grants and conduct research, and getting a Ph.D. requires four to six years of successful postgraduate academic performance with a high level of writing skills and a large vocabulary — which are essentially attributes measured by verbal intelligence tests. Furthermore, most science Ph.D. programs demand mathematical and logical thinking, which are measured by other components of intelligence tests. When college graduates apply to graduate school they have to take such tests as the Graduate Record Examination (GRE), which measure these abilities, and only the high-scoring students are accepted into science graduate programs. Thus, from this perspective, it is not surprising that scientists generally have IQ scores of 110 to 120 or above: without the ability to achieve such scores, it is unlikely they would have ever had the chance to become scientists in the first place.

One could also speculate that there are certain minimum "talent" requirements for such things as playing sports or painting, so that people who fall below those requirements would find it difficult or impossible to become highly skilled in those areas. But, outside of some very basic physical traits, such as height and body size in sports, we have no solid evidence that such minimum requirements exist.

We do know — and this is important — that among those people who have practiced enough and have reached a certain level of skill in their chosen field, there is no evidence that any genetically determined abilities play a role in deciding who will be among the best. Once you get to the top, it isn't natural talent that makes the difference, at least not "talent" in the way it is usually understood as an innate ability to excel at a particular activity.

I believe this explains why it is so difficult to predict who will rise to the top of any given field. If some sort of innate ability were playing a role in deciding who eventually becomes the best in a particular area, it would be much easier to spot those future champions early in their careers. If, for instance, the best professional football players were the ones who were born with some sort of gift for football, then that gift should certainly be apparent by the time they're in college, at which point they have generally been playing football for half a dozen years or more. But in reality, no one has figured out how to look at college football players and figure out which will be the best and which will be duds. In 2007, quarterback JaMarcus Russell of Louisiana State University was chosen first overall in the NFL draft; he was a complete bust and was out of football within three years. By contrast, Tom Brady was picked in the sixth round of the 2000 draft — after 198 other players — and he developed into one of the best quarterbacks ever.

A 2012 study of tennis players looked at the success and rankings of junior tennis players — that is, younger players who are working and competing to become professionals — and compared that with their success after turning pro. There was no relationship. If differences in innate talent were playing a role in determining the best professional tennis players, you'd think those differences would have been noticeable during their junior tennis years as well, but they were not.

The bottom line is that no one has ever managed to figure out how to identify people with "innate talent." No one has ever found a gene variant that predicts superior performance in one area or another, and no one has ever come up with a way to, say, test young children and identify which among them will become the best athletes or the best mathematicians or the best doctors or the best musicians.

There is a simple reason for this. If there are indeed genetic differences that play a role in influencing how well someone performs (beyond the initial stages when someone is just learning a skill), they aren't likely to be something that affects the relevant skills directly — a "music gene" or a "chess gene" or a "math gene." No, I suspect that such genetic differences — if they exist — are most likely to manifest themselves through the necessary practice and efforts that go into developing a skill. Perhaps, for example, some children are born with a suite of genes that cause them to get more pleasure from drawing or from making music. Then those children will be more likely to draw or to make music than other children. If they're put in art classes or music classes, they're likely to spend more time practicing because it is more fun for them. They carry their sketchpads or guitars with them wherever they go. And over time these children will become better artists or better musicians than their peers — not because they are innately more talented in the sense that they have some genes for musical or artistic ability, but because something — perhaps genetic — pushed them to practice and thus develop their skills to a greater degree than their peers.

Research on the development of vocabulary in very young children has shown that such factors as the child's temperament and ability to pay attention to a parent influence the size of vocabulary the child will build. Most of a young child's vocabulary development comes through interaction with a parent or other caregiver, and studies have shown that children with a temperament that encourages social interaction end up developing better language skills. Similarly — and more in line with the sorts of factors that may play a role in acquiring skills with practice — nine-month-old infants who paid more attention to a parent as that parent was reading a book and pointing to the pictures in the book grew up to have a much better vocabulary at five years of age than infants who paid less attention.

It is possible to imagine a number of genetically based differences of this sort. Some people might, for instance, be naturally able to focus more intently and for longer periods of time than others; since deliberate practice depends on being able to focus in this way, these people might be naturally able to practice more effectively than others and thus benefit more from their practice. One could even imagine differences in how the brain responds to challenges so that practice would be more effective in some people than in others in building new brain structures and mental capacity.

Much of this remains speculative at this point. But since we know that practice is the single most important factor in determining a person's ultimate achievement in a given domain, it makes sense that if genes do play a role, their role would play out through shaping how likely a person is to engage in deliberate practice or how effective that practice is likely to be. Seeing it in this way puts genetic differences in a completely different light.

THE DARK SIDE OF BELIEVING IN INNATE TALENT

In this chapter I've discussed the roles that practice and innate talent play in the development of expert performers, and I've argued that while innate characteristics may influence performance among those who are just learning a new skill or ability, the degree and the effectiveness of training plays a more significant role in determining who excels among those who have worked to develop a skill. This is because, ultimately, the body's and the brain's natural ability to adapt in the face of challenges outweighs any genetic differences that may, in the beginning, give some people an advantage. So I believe that it's much more important to understand how and why particular types of practice lead to improvement than it is to go looking for genetic differences between people.

But there is, I believe, an even more urgent reason to emphasize the role of practice over that of innate differences, and that is the danger of the self-fulfilling prophecy.

When people assume that talent plays a major, even determining, role in how accomplished a person can become, that assumption points one toward certain decisions and actions. If you assume that people who are not innately gifted are never going to be good at something, then the children who don't excel at something right away are encouraged to try something else. The clumsy ones are pushed away from sports, the ones who can't carry a tune right away are told they should try something other than music, and the ones who don't immediately get comfortable with numbers are told they are no good at math. And, no surprise, the predictions come true: the girl who was told to forget about sports never becomes any good at hitting a tennis ball or kicking a soccer ball; the boy who was told he was tone-deaf never learns to play a musical instrument or to sing well; and the children who were told they were no good at math grow up believing it. The prophecy becomes self-fulfilling.

On the flip side, of course, the children who get more attention and praise from their teachers and coaches and more support and encouragement from their parents do end up developing their abilities to a much greater degree than the ones who were told never to try — thus convincing everyone that their initial appraisals were correct. Again, self-fulfilling.

Malcolm Gladwell told a story in his book *Outliers* — a story that others had told before him, but it was Gladwell's telling that got the most attention — of how there are many more Canadian professional hockey players born in the months of January through March than born in October through December. Is there something magical about being born in these months that grants extra talent for hockey to babies lucky enough to be born then? No. What happens is that there is a cutoff for playing youth hockey in Canada — you must be a certain age by December 31 of the previous year — and the children born in the first three months of the year are the oldest players in each class of players. When children start playing hockey at around age four or five, the advantage that older kids can have over younger ones is striking. Kids with an age advantage of close to a year will generally be taller, heavier, and somewhat more coordinated and mentally mature, and they may have had one more hockey season to develop their hockey skills, so they are likely to be better at hockey than the younger players in their age group. But those age-related physical differences get smaller and smaller as the hockey players get older, and they have pretty much disappeared by the time the players reach adulthood. So the age-related advantage must have its roots in childhood, when the physical differences still exist.

The obvious explanation for the age effect is that it starts with the coaches, who are searching for the most talented players, beginning at the very earliest ages. Coaches can't really tell how old the various child hockey players are; all they can see is who is doing better and thus, by inference, who appears to be more talented. Many coaches will tend to treat the more "talented" players with more praise and better instruction and to give these players more opportunities to play in games. And these players will be viewed as more talented not just by the coach but also by the other players. Furthermore, these players might be more willing to practice more because they are told that they have the promise of playing at very high levels, even professionally. The results of all this are striking—and not just in hockey. For example, one study found that among thirteen-year-old soccer players, more than 90 percent of the ones who were nominated as the best had been born in the first six months of the year.

The advantage among hockey players does seem to taper off somewhat once the players make it into the major leagues—perhaps because the younger ones who have managed to hang around have learned to work harder at their practicing and thus end up outshining many of the ones who are six months older—but there is no doubt that being born in January through March is an advantage to any Canadian boy who wants to play hockey.

Now suppose that the same thing happened with chess. Suppose there was some group of people who selected beginning chess players for some chess program according to what seemed to be their "innate talent." They would teach a group of youngsters how to play and then, after three or six months had passed, look to see who were the best. We know what would happen. On average, the kids with higher IQs would have an easier time in the beginning learning the moves and would be selected for further training and grooming; the others would not be offered a spot in the program. The end result would be a collection of chess players with much higher than average IQs. But we know that in the real world there are many grandmasters who don't score particularly well on IQ tests — so we would have missed the contributions of all of those people who could become great chess players.

And now suppose that we're not talking about a chess program but rather math as it is taught in most schools. No one has done studies with math that match the ones done with chess, but suppose for a moment that a similar thing is true — that is, that children with a higher spatial intelligence can learn to do basic math more quickly than others. Recent research has shown that children who have had experience playing linear board games with counting steps before they start school will do better in math once they are in school. And there are likely many other ways that certain preschool experiences will help children perform better in math later on. Most teachers, however, are not familiar with this possibility, so when some kids "get" math more quickly than others, they're generally assumed to be gifted at math while the others aren't. Then the "gifted" ones get more encouragement, more training, and so on, and, sure enough, after a year or so they're much better at math than the others, and this advantage propagates through the school years. Since there are a number of careers, like engineering or physics, that require math courses in college, the students who have been judged to have no talent for math find these careers closed to them. But if math works the same way as chess, then we have lost a whole collection of children who might eventually have become quite accomplished in these areas if only they hadn't been labeled as "no good at math" in the very beginning.

This is the dark side of believing in innate talent. It can beget a tendency to assume that some people have a talent for something and others don't and that you can tell the difference early on. If you believe that, you encourage and support the "talented" ones and discourage the rest, creating the self-fulfilling prophecy. It is human nature to want to put effort — time, money, teaching, encouragement, support — where it will do the most good and also to try to protect kids from disappointment. There is usually nothing nefarious going on here, but the results can be incredibly damaging. The best way to avoid this is to recognize the potential in all of us — and work to find ways to develop it.

9

Where Do We Go from Here?

CALL IT A GLIMPSE. For one week a group of students enrolled in a traditional freshman physics class got a look at what the future of learning physics might look like. It was just one section on electromagnetic waves that was taught toward the end of a two-semester course, but in that one section the results were almost magical. The students who were taught the material with a method inspired by the principles of deliberate practice learned more than twice as much as those students taught with the traditional approach. By one measure it was the largest effect ever seen in an educational intervention.

This glimpse came courtesy of three researchers associated with the University of British Columbia: Louis Deslauriers, Ellen Schelew, and Carl Wieman. Wieman, who won the Nobel Prize in Physics in 2001, has made a second career out of working to improve undergraduate science education. Using part of his Nobel Prize winnings, in 2002 he created the Physics Education Technology Project at the University of Colorado, and later he established the Carl Wieman Science Education Initiative at the University of British Columbia. In all this he has been driven by the conviction that there is a better way to teach science than the traditional fifty-minute classroom lectures. And this is what he and his two colleagues set out to demonstrate in that bastion of traditional teaching, the freshman physics course.

The class at UBC had 850 students in three sections. It was a hardcore physics course, aimed at first-year engineering majors, with the physics concepts taught in terms of calculus and the students expected to learn how to solve math-intensive problems. The professors were well regarded for their teaching skills, with years of experience teaching this particular course and good scores on their student evaluations. Their method of instruction was relatively standard: three fifty-minute PowerPoint lectures a week given in a large lecture hall, weekly homework assignments, and tutorial sessions where the students would solve problems under the eye of a teaching assistant.

Wieman and his colleagues chose two of the course's sections, each with about 270 students, to serve as their testing ground. For the twelfth week of the second semester, one of these sections would continue with instruction as usual, while the other would be presented with a completely different way to learn about electromagnetic waves. The students in the two sections were about as alike as they could possibly be: the average scores on the two midterm tests the students had taken up to that point were identical between the two classes; the average class scores on two standardized tests of physics knowledge given during week eleven were identical; the class attendance rates during weeks ten and eleven were identical; and the assessed levels of engagement during weeks ten and eleven were identical for the two classes. In short, up to that point the two classes had been essentially identical in their classroom behavior and how well they were learning about physics. That was about to change.

In the twelfth week, as the instructor of one section continued on as usual, the instructor in the second section was replaced with Wieman's two colleagues, Deslauriers and Schelew. Deslauriers served as the main instructor and Schelew as his assistant. Neither of them had ever been in charge of a class before. Deslauriers, a postdoctoral student, had received some training in effective teaching methods and, in particular, the teaching of physics during his time at the Carl Wieman Science Education Initiative. Schelew was a physics graduate student who had taken a seminar in physics education. Both had spent some time as teaching assistants. But together they had far less experience in the classroom than the instructor who was continuing to teach the other section during the week of the trial.

What Deslauriers and Schelew did have was a new approach to teaching physics that Wieman and others had developed by applying the principles of deliberate practice. For one week they had the students in their section follow a very different pattern than in the traditional class. Before each class they were expected to read assigned sections — generally just three or four pages long — from their physics text and then complete a short online true/false test about the reading. The idea was to make them familiar with the concepts that would be worked on in class before they ever came to class. (To even things out, the students in the traditional class were also asked to do preclass reading during this one week. It was the only change made in how the traditional class was taught during that week.)

In the deliberate-practice class the goal was not to feed information to the students but rather to get them to practice thinking like physicists. To do that, Deslauriers would first have the students divide up into small groups and then pose a "clicker question," that is, a question that the students answered electronically, with the answers sent automatically to the instructor. The questions were chosen to get the students in the class thinking about concepts that typically give first-year physics students difficulty. The students would talk about each question within their small groups, send in their answers, and then Deslauriers would display the results and talk about them, answering any questions that the students might have. The discussions got the students thinking about the concepts, drawing connections, and often moving beyond the specific clicker question they'd been asked. Several clicker questions were asked during the course of the class, and sometimes Deslauriers might have the student groups discuss a question a second time, after he had offered some thoughts for them to ponder. Sometimes he would offer a mini-lecture if it seemed that the students were having difficulty with a particular idea. Each class also included an "active learning task" in which the students in each group considered a question and then individually wrote their answers and submitted them, after which Deslauriers would again answer questions and address misconceptions. During the class Schelew would walk around among the groups, answering questions, listening to the discussions, and identifying problem areas.

The students were much more active participants in this class than in the traditionally taught class. This was demonstrated by the measures of engagement that Wieman's group used. Although there was no difference in engagement between the two groups during weeks ten and eleven, during week twelve the engagement in the class taught by Deslauriers was nearly double what it was in the traditional class. But it was more than just engagement. The students in the Deslauriers class were getting immediate feedback on their understanding of the various concepts, with both fellow students and the instructors helping clear up any confusion. And both the clicker questions and the active learning tasks were designed to get the students thinking like physicists—to first understand the question in the proper way, then figure out which concepts are applicable, and then reason from those concepts to an answer. (The instructor in the traditional class observed Deslauriers's class before teaching his own and chose to use most of the same clicker questions in his own class, but he did not use them to begin discussions, only to show the class how many students had gotten each answer correct.)

At the end of week twelve, the students in both class sections were given a multiple-choice clicker test to see how well they had learned the material. Deslauriers and the instructor in the traditional class had worked together to develop a test that they and the instructor of a third section all agreed was a good measure of the learning objectives for that week. The test questions were very standard. Indeed, most of them were clicker questions that had been used for a physics class at another university, sometimes with small modifications.

The average score of the students in the traditional section was 41 percent; the average in Deslauriers's class was 74 percent. That is obviously a big difference, but given that random guessing would have produced a score of 23 percent, when you do the math it turns out that the students in the traditional class, on average, knew the right answer on only about 24 percent of the questions, compared with an average of about 66 percent in the class designed to apply the principles of deliberate practice. That is a huge difference. The students in the deliberate practice class got more than 2.5 times as many right answers as those in the other class.

Wieman and his colleagues expressed the difference in another way, using a statistical term known as "the effect size." In these terms the difference between the performances of the two classes was 2.5 standard deviations. For the sake of comparison, other new teaching methods in science and engineering classrooms generally have effect sizes less than 1.0, and the largest effect size observed for an educational intervention before this had been 2.0 — which was accomplished with the use of trained personal tutors. Wieman got to 2.5 with a graduate student and a postdoc who had never taught a class before.

THE PROMISE OF DELIBERATE PRACTICE

Wieman's achievement is tremendously exciting. It suggests that by modifying traditional teaching approaches to reflect the insights of deliberate practice, we might dramatically improve the effectiveness of teaching in various fields. So where do we start?

One place would be with the development of world-class athletes, musicians, and other expert performers. I've always hoped that the work I've done in understanding deliberate practice would prove useful to these performers and their coaches. After all, not only are they the people who are most interested in finding ways to improve performance, but they're also the ones from whom I have learned the most in my research. And, indeed, I believe that there is much that expert performers and aspiring expert performers can do to improve their training.

For instance, it has always been surprising to me when I talk to full-time athletes and their coaches how many of them have never taken the time to identify those aspects of performance that they would like to improve and then design training methods aimed specifically at those things. In reality, much of the training that athletes do — especially athletes in team sports — is carried out in groups with no attempt to figure out what each individual should be focusing on.

Furthermore, very little has been done to learn about the mental representations that successful athletes use. The ideal approach to fixing this would be to have athletes verbally report their thinking while they are performing, which would make it possible for researchers, coaches, or perhaps even the athletes themselves to design training tasks to improve their representations of game situations, in the same way that we described in chapter 3. There are, of course, some elite athletes who develop effective representations by themselves, but most of these top players are not even aware of how their thinking differs from those less accomplished. And the converse is certainly often true as well — that the less accomplished athletes don't understand how much weaker their mental representations are than those of the best in their sport.

For example, over the past few years I've spoken with coaches in a variety of sports, including Chip Kelly, head coach of the Philadelphia Eagles of the National Football League. These coaches are generally eager to learn how deliberate practice can improve the performance of their athletes. In a group meeting I had with all of the Eagles coaches in the spring of 2014, we discussed how all the great players seem to be aware of what the relevant team and opposing players were doing so they could discuss it after a training session or a game. However, I found that even those coaches who recognized the importance of effective mental representations did little to help the less elite players improve their representations; instead, they generally found it easier to pick players who had already acquired effective mental representations and then provide them with additional training to further improve those representations.

During a visit to the Manchester City Football Club in England in 2011 (before that team won the Football Association Challenge Cup) I discussed similar issues. The coaches there were more receptive to talking about how to train representations because they trained young players, several of whom would eventually be allowed to play on an adult team during regular matches.

I have also been working with Rod Havriluk, a swimming coach and president of the International Society of Swimming Coaching, to use insights from deliberate practice to improve swimming instruction. Rod and I have found that there is hardly any individualized coaching of — or deliberate practice by — swimmers at the lower and middle levels.

Given how little work has been done to apply the principles of deliberate practice to the development of expert performers, particularly athletes, it is clear there is great potential for improvement by focusing on individualized training and the assessment of the athletes' mental representations. And I will continue to work with coaches, trainers, and athletes to help them use deliberate practice more effectively.

But the greatest potential benefits from deliberate practice, I believe, lie elsewhere. After all, the top performers in the various highly specialized and highly competitive fields — the professional athletes, the world-class musicians, the chess grandmasters, and so on — make up just a tiny fraction of the world's population, and while it is a very visible and entertaining fraction, it will make a relatively small difference to the rest of the world if these few people get marginally better at what they do. There are other areas where many more people can be helped and where the improvements can be much larger because the training in those areas is even further away from what deliberate practice tells us is the ideal.

Education is one of those areas. Education touches everyone, and there are a number of ways that deliberate practice could revolutionize how people learn.

The first is pedagogical. How do students learn best? Deliberate practice has a great deal to say about that question.

Let's take a closer look at that UBC physics class to see how the principles of deliberate practice can be applied to help students learn faster and better than they do with traditional approaches. The first thing that Wieman and his colleagues did in designing the class was to speak with the traditional instructors to determine exactly what the students should be able to do once they finished the section.

As we discussed in chapter 5, a major difference between the deliberate-practice approach and the traditional approach to learning lies with the emphasis placed on skills versus knowledge — what you can do versus what you know. Deliberate practice is all about the skills. You pick up the necessary knowledge in order to develop the skills; knowledge should never be an end in itself. Nonetheless, deliberate practice results in students picking up quite a lot of knowledge along the way.

If you teach a student facts, concepts, and rules, those things go into long-term memory as individual pieces, and if a student then wishes to do something with them — use them to solve a problem, reason with them to answer a question, or organize and analyze them to come up with a theme or a hypothesis — the limitations of attention and short-term memory kick in. The student must keep all of these different, unconnected pieces in mind while working with them toward a solution. However, if this information is assimilated as part of building mental representations aimed at doing something, the individual pieces become part of an interconnected pattern that provides context and meaning to the information, making it easier to work with. As we saw in chapter 3, you don't build mental representations by thinking about something; you build them by trying to do something, failing, revising, and trying again, over and over. When you're done, not only have you developed an effective mental representation for the skill you were developing, but you have also absorbed a great deal of information connected with that skill.

When preparing a lesson plan, determining what a student should be able to do is far more effective than determining what that student should know. It then turns out that the knowing part comes along for the ride.

Once Wieman and his colleagues had put together a list of what things their students should be able to do, they transformed it into a collection of specific learning objectives. Again, this is a classic deliberate-practice approach: when teaching a skill, break the lesson into a series of steps that the student can master one at a time, building from one to the next to reach the ultimate objective. While this sounds very similar to the scaffolding approach used in traditional education, it differs crucially in its focus on understanding the necessary mental representations at each step of the way and making sure that the student has developed the appropriate representations before moving to the next step. This seems to have been, for example, the crucial ingredient in the success of the Jump Math program described in the last chapter: the program carefully delineates which representations are necessary for the development of a particular math skill and then teaches in a way that builds those representations in the students.

Generally speaking, in almost any area of education the most useful learning objectives will be those that help students develop effective mental representations. In physics, for example, it is always possible to teach students how to solve particular equations and how to decide which equations should be applied in which situations, but that's not the most important part of what physicists know. Research comparing physics experts with physics students has found that while the students may sometimes be almost as good as the experts at solving quantitative problems — that is, problems involving numbers that can be solved by applying the right equation — the students are far behind the experts in their ability to solve qualitative problems, or problems that involve concepts but no numbers, something like, Why is it hot in summer and cold in the winter? Answering a question like that requires less a command of numbers than it does a clear understanding of the concepts that underlie particular events or processes — that is, good mental representations.

Most people — with the exception of science teachers — cannot correctly explain what causes the changing seasons, even though it is taught in science classes as early as elementary school. An amusing video taken at a Harvard University commencement shows a string of recent graduates confidently explaining that the seasons result from the Earth being closer to the sun in summer and farther away in winter. This is completely wrong, of course, since when it is summer in the Northern Hemisphere, it is winter in the Southern Hemisphere. The real cause of the seasons is the tilt of the Earth on its axis. But the point here is not the ignorance of Harvard graduates but rather that so little of science education gives students the basic mental representations they need to think clearly about physical phenomena rather than teaching them simply to plug numbers into an equation.

To help the physics students in their class develop such mental representations, Wieman and his coworkers developed sets of clicker questions and learning tasks that would help them reach the learning objectives the instructors had previously identified. The clicker questions and tasks were chosen to trigger discussions that would lead the students to grapple with and apply the concepts they were learning and, ultimately, to use those concepts to answer the questions and solve the tasks.

The questions and tasks were also designed to push the students outside their comfort zones — to ask them questions whose answers they'd have to struggle for — but not so far outside their comfort zones that they wouldn't know how to start answering them. Wieman and his colleagues pretested the clicker questions and learning tasks on a couple of student volunteers who were enrolled in the course. They gave these students the questions and the learning tasks and then had them think aloud as they reasoned their way toward the answers. Based on what the researchers heard during the think-aloud sessions, they modified the questions and tasks, with a specific emphasis on avoiding misunderstandings and questions that were too difficult for the students to deal with. Then they went through a second round of testing with another volunteer, sharpening the questions and learning tasks even more.

Finally, the classes were structured so that the students would have the opportunity to deal with the various concepts over and over again, getting feedback that identified their mistakes and showed how to correct them. Some of the feedback came from fellow students in the discussion groups and some from the instructors, but the important thing was that the students were getting immediate responses that told them when they were doing something wrong and how to fix it.

The redesigned physics class at the University of British Columbia offers a road map for redesigning instruction according to deliberate-practice principles: Begin by identifying what students should learn how to do. The objectives should be skills, not knowledge. In figuring out the particular way students should learn a skill, examine how the experts do it. In particular, understand as much as possible about the mental representations that experts use, and teach the skill so as to help students develop similar mental representations. This will involve teaching the skill step by step, with each step designed to keep students out of their comfort zone but not so far out that they cannot master that step. Then give plenty of repetition and feedback; the regular cycle of try, fail, get feedback, try again, and so on is how the students will build their mental representations.

At the University of British Columbia, the success of Wieman's deliberate practice–based approach to teaching physics has led many other professors there to follow suit. According to an article in *Science* magazine, in the years after the experiment deliberate-practice methods were adopted in nearly one hundred science and mathematics classes there with a total enrollment of more than thirty thousand students. Since math and science professors have traditionally been very resistant to changing their teaching methods, this says a great deal about the quality of Wieman's findings.

Redesigning teaching methods using deliberate practice could dramatically increase how quickly and how well students learn — as the almost unbelievable improvements in Wieman's students indicates — but it will require not only a change in mindset among educators but much more research into the minds of experts. We're only just starting to understand the types of mental representations that experts use and how to develop these representations with deliberate practice. There is much more to do.

In addition to more effective teaching methods, there are other, less obvious ways that deliberate practice can be applied to education. In particular I think there would be tremendous value in helping children and, especially, adolescents develop detailed mental representations in at least one area, for reasons that we will discuss below. This is not a goal of the current educational system, and generally the only students who do develop such representations are those who are pursuing some skill outside of school — playing a sport or a musical instrument, for instance — and even then the students do not really understand what they're doing or recognize that their representations are part of a larger phenomenon that stretches across fields.

One benefit that a young student — or anyone, really — gets from developing mental representations is the freedom to begin exploring that skill on his or her own. In music, having clear representations of what musical pieces sound like, how the different sections of a piece fit together to create the whole, and how variations in one's playing can affect the sound allows student musicians to play music for themselves or for others and to improvise and explore on their instruments. They no longer need a teacher to lead them down every path; they can head down some paths on their own.

Something similar is true for academic subjects. Students who develop mental representations can go on to generate their own scientific experiments or to write their own books — and research has shown that many successful scientists and authors started their careers at a young age in just this way. The best way to help students develop their skills and mental representations in an area is to give them models they can replicate and learn from, just as Benjamin Franklin did when he improved his writing by reproducing articles from *The Spectator.* They need to try and fail — but with ready access to models that show what success looks like.

Having students create mental representations in one area helps them understand exactly what it takes to be successful not only in that area but in others as well. Most people, even adults, have never attained a level of performance in any field that is sufficient to show them the true power of mental representations to plan, execute, and evaluate their performance in the way that expert performers do. And thus they never really understand what it takes to reach this level — not just the time it takes, but the high-quality practice. Once they do understand what is necessary to get there in one area, they understand, at least in principle, what it takes in other areas. That is why experts in one field can often appreciate those in other fields. A research physicist may better understand what it takes to become a skilled violinist, if only in general terms, and a ballerina may better understand the sacrifice it takes to become a skilled painter.

Our schools should give all students such an experience in some domain. Only then will they understand what is possible and also what it takes to make it happen.

HOMO EXERCENS

In the introduction to this book I spoke of how deliberate practice can revolutionize our thinking about human potential. I do not see this as an exaggeration or overstatement. That revolution starts when we realize that the best among us in various areas do not occupy that perch because they were born with some innate talent but rather because they have developed their abilities through years of practice, taking advantage of the adaptability of the human body and brain.

But that realization is not enough. We need to give people the tools they need to harness that adaptability and take control of their own potential. Spreading the word about deliberate practice — as I am doing with this book — is part of that, but many of the necessary tools are still undeveloped. In most fields we still don't know exactly what distinguishes experts from everyone else. Nor do we have many details about the experts' mental representations. We need to map out the various factors that make up an expert over his or her entire lifespan in order to provide direction for other people who want to develop expertise.

Even before we have a comprehensive road map, however, we can get a good start down that road. As I mentioned above, we can help students develop expertise and effective mental representations in at least one area so that they can learn about expertise itself — what produces it and how accessible it is to everyone. And, as we discussed in chapter 6, developing a skill through deliberate practice can increase the motivation for further improvement because of the positive feedback one gets from possessing that skill. If we can show students that they have the power to develop a skill of their choice and that, while it is not easy, it has many rewards that will make it worthwhile, we make it much more likely that they will use deliberate practice to develop various skills over their lifetimes.

Over time, then, by learning more about what goes into expert performance in various fields and by creating a generation of students primed to take advantage of that, we could produce a new world, one in which most people understand deliberate practice and use it to enrich their lives and their children's.

What kind of world would that be? To begin with, it would contain far more experts in far more fields than we have today. The societal implications of this would be enormous. Imagine a world in which doctors, teachers, engineers, pilots, computer programmers, and many other professionals honed their skills in the same way that violinists, chess players, and ballerinas do now. Imagine a world in which 50 percent of the people in these professions learn to perform at the level that only the top 5 percent manage today. What would that mean for our health care, our educational system, our technology?

The personal benefits could be tremendous as well. I have spoken very little of this here, but expert performers get great satisfaction and pleasure from exercising their abilities, and they feel a tremendous sense of personal accomplishment from pushing themselves to develop new skills, particularly skills that are on the very edges of their fields. It is as if they are on a constantly stimulating journey where boredom is never a problem because there are always new challenges and opportunities. And those experts whose skills relate to some sort of performance — the musicians, dancers, gymnasts, and so forth — report getting great pleasure from performing in public. When everything goes well they experience a level of effortlessness similar in many ways to the psychological state of "flow" popularized by Mihaly Csikszentmihalyi. This gives them a precious "high" that few people other than experts ever experience.

One of the most exciting times in my life was when I worked with Herb Simon and he received the Nobel Prize. Everyone in our group had this sense of being at the frontier of our area of science and feeling really lucky to be there. I imagine it must have been the same sort of excitement that the Impressionists felt as they were working to revolutionize art.

Even those who don't reach the frontiers of a field can still enjoy the challenge of taking control of their own lives and improving their abilities. A world in which deliberate practice is a normal part of life would be one in which people had more volition and satisfaction.

And I would argue that we humans are most human when we're improving ourselves. We, unlike any other animal, can consciously change ourselves, to improve ourselves in ways we choose. This distinguishes us from every other species alive today and, as far as we know, from every other species that has ever lived.

The classic conception of human nature is captured in the name we gave ourselves as a species, *Homo sapiens.* Our distant ancestors included *Homo erectus,* or "upright man," because the species could walk upright, and *Homo habilis,* the "handy man," so named because the species was at one time thought to be the earliest humans to have made and used stone tools. We call ourselves "knowing man" because we see ourselves as distinguished from our ancestors by our vast amount of knowledge. But perhaps a better way to see ourselves would be as *Homo exercens,* or "practicing man," the species that takes control of its life through practice and makes of itself what it will.

It is quite possible that this new understanding couldn't have come at a better time. Thanks to technology our world is changing at an ever-increasing pace. Two hundred years ago a person could learn a craft or trade and be fairly certain that that education would suffice for a lifetime. People born in my generation grew up thinking the same way: get an education, get a job, and you'll be set until you retire. That has changed in my lifetime. Many jobs that existed forty years ago have disappeared or else have changed so much as to be almost unrecognizable. And people coming into the work force today should expect to change careers two or three times during their working lives. As for the children being born today, no one knows, but I think it's safe to say that the changes won't be slowing down.

How do we as a society prepare for that? In the future most people will have no choice but to continuously learn new skills, so it will be essential to train students and adults about how to learn efficiently. With the technological revolution there are new opportunities to make teaching more effective. It is possible, for example, to videotape real-world experiences of doctors, athletes, and teachers and create libraries and learning centers where students could be trained in such a way that avoids having to learn on the job and risk the welfare of patients, students, and clients.

We need to start now. For adults who are already in the work world, we need to develop better training techniques — based on the principles of deliberate practice and aimed at creating more effective mental representations — that not only will help them improve the skills they use in their current jobs but that will enable them to develop new skills for new jobs. And we need to get the message out: you can take charge of your own potential.

But it is the coming generations who have the most to gain. The most important gifts we can give our children are the confidence in their ability to remake themselves again and again and the tools with which to do that job. They will need to see firsthand — through their own experiences of developing abilities they thought were beyond them — that they control their abilities and are not held hostage by some antiquated idea of natural talent. And they will need to be given the knowledge and support to improve themselves in whatever ways they choose.

Ultimately, it may be that the only answer to a world in which rapidly improving technologies are constantly changing the conditions under which we work, play, and live will be to create a society of people who recognize that they can control their development and understand how to do it. This new world of *Homo exercens* may well be the ultimate result of what we have learned and will learn about deliberate practice and about the power it gives us to take our future into our own hands.

Acknowledgments

The research that I have completed was possible because of the factors that I have described in this book. My parents provided me with a safe environment, where I was encouraged to pursue any type of project as long as I was willing to do whatever it took. At the University of Stockholm in Sweden I was supervised by Professor Gunnar Goude, who was willing to encourage and support my interest in research on thinking while his interests were focused on research on animals, thus forcing me to think independently. Herbert Simon and Bill Chase at Carnegie Mellon showed me how to find and study important problems and helped me get a job as a professor in psychology in the University of Colorado in the United States. Paul Baltes at the Max Planck Institute for Human Development in Berlin gave me the opportunity and resources to conduct the research on the music students in collaboration with Ralf Krampe and Clemens Tesch-Römer. I want to thank my many students and postdoctoral fellows and other collaborators, especially Andreas Lehmann. I want to thank the many experts and participants who shared their thinking and allowed the study of their performances. Finally I want to thank and acknowledge the participants in the long-term training studies, in particular, Steve Faloon, Dario Donatelli, John Conrad, and Rajan Mahadevan.

My research has been supported by grants from the Office of Naval Research as principal investigator (N00014-84-K-0250) and co-principal investigator (N00014-04-1-0588, N00014-05-1-0785, N00014-07-1-0189), a grant from the U.S. Army Research Institute to the University of Colorado as a principal investigator (CU-1530638), grants from the Max Planck Society as principal investigator, a grant from the U.S. Soccer Foundation as a co-principal investigator (FSU Research Foundation grant 1 1520 0006), and research funds from the Conradi Eminent Scholar Endowment at the Florida State Foundation as principal investigator.

—Anders Ericsson

I wish to thank Thomas Joiner of the Florida State University Department of Psychology for introducing me to Anders Ericsson many years ago, without which this book would have never happened, and to thank Anders himself, one of the most generous people with ideas and insights whom I have ever met. What I learned about him from deliberate practice has immeasurably enriched my life, and that would have been true even if we had never written the book. I'd also like to thank Art Turock for providing some fascinating examples of how deliberate practice can be applied in the business world.

Finally, my greatest and most heartfelt thanks go to my wife, Deanne Laura Pool, for her many, many contributions to this book. She served as an idea generator, a sounding board, a discerning first reader, and an extraordinary editor throughout the entire (very long) process of writing this book. She shaped my thinking on the subject in countless ways, large and small, by discussing ideas, asking probing questions, offering thoughtful suggestions, spotting weaknesses, and pointing out strengths. A writer herself, she was responsible for making this book much better thought-out and much better written than it otherwise would have been. While her name may not be on the cover, her fingerprints are all over the book.

— Robert Pool

Both of us wish to thank Elyse Cheney and Alex Jacobs for all their support and efforts to help us shape a book proposal and, eventually, the book itself, to be of interest to as many readers as possible. We are also very grateful to our editor, Eamon Dolan, for the thoughtful and challenging issues and ideas that he proposed, all of which greatly improved the structure of our arguments and of the book.

Notes

Introduction: The Gift

page

xii *rather breathless letter:* The letter describing Mozart's perfect pitch can be found in Otto Erich Deutsch, *Mozart: A Documentary Biography,* 3rd ed. (London: Simon and Schuster, 1990), 21. See also Diana Deutsch, "Absolute pitch," in *The Psychology of Music,* ed. Diana Deutsch, 3rd ed. (San Diego: Elsevier, 1990), 141–182.

xiii *only about one in every ten thousand people:* See, for example, William Lee Adams, "The mysteries of perfect pitch," *Psychology Today,* July 1, 2006, https://www.psychologytoday.com/articles/200607/the-mysteries-perfect-pitch (accessed February 25, 2015).

around three to five years old: Robert J. Zatorre, "Absolute pitch: A model for understanding the influence of genes and development on neural and cognitive function," *Nature Neuroscience* 6, no. 7 (2003): 692–695. See also Siamak Baharloo, Paul A. Johnston, Susan K. Service, Jane Gitschier, and Nelson B. Freimer, "Absolute pitch: An approach for identification of genetic and nongenetic components," *American Journal of Human Genetics* 62 (1998): 224–231.

xiv *other ethnicities to have perfect pitch:* Diana Deutsch, Kevin Dooley, Trevor Henthorn, and Brian Head, "Absolute pitch among students in an American music conservatory: Association with tone language fluency," *Journal of the Acoustical Society of America* 125 (2009): 2398–2403.

pretty much what we knew: My own review of evidence for the acquired nature of perfect pitch is summarized in K. Anders Ericsson and Irene Faivre, "What's exceptional about exceptional abilities?" in *The Exceptional Brain: Neuropsychology of Talent and Special Abilities,* ed. Loraine K. Obler and Deborah Fein (New York: Guilford, 1988), 436–473.

scientific journal Psychology of Music*:* Ayako Sakakibara, "A longitudinal study of the process of acquiring absolute pitch: A practical report of training with the 'chord identification method,'" *Psychology of Music* 42, no. 1 (2014): 86–111.

xv *notes played on the piano:* Two of the twenty-four children dropped out over the course of the training, but their departures had nothing to do with how well their training was going. All twenty-two children who finished the training exhibited perfect pitch.

professional adult musicians: Deutsch, *Mozart,* 21.

violin, the keyboard, and more: Stanley Sadie, *Mozart: The Early Years, 1756–1781* (New York: W. W. Norton, 2006), 18.

xviii *an artistic gymnast at the international level:* The average adult height of international-level gymnasts is five feet two inches, with the upper bound being five feet seven inches. Neoklis A. Georgopoulos, Anastasia Theodoropoulou, Nikolaos D. Roupas, et al., "Growth velocity and final height in elite female rhythmic and artistic gymnasts," *Hormones* 11, no. 1 (2012): 61–69.

xix *assumed he was born with it:* Jackie MacMullan, "Preparation is key to Ray Allen's 3's," *ESPN Magazine,* February 11, 2011, http://sports.espn.go.com/boston/nba/columns/story?columnist=macmullan_jackie&id=6106450 (accessed March 30, 2015).

xx *motivation, and effort:* See, for example, Malcolm Gladwell, *Outliers: The Story of Success* (New York: Little, Brown, 2008); David Shenk, *The Genius in All of Us: Why Everything You've Been Told About Genetics, Talent, and IQ Is Wrong* (New York: Doubleday, 2010); Carol Dweck, *Mindset: The New Psychology of Success* (New York: Random House, 2006). There are many more: K. Anders Ericsson and Jacqui Smith, eds., *Toward a General Theory of Expertise: Prospects and Limits* (Cambridge, UK: Cambridge University Press, 1991); K. Anders Ericsson, ed., *The Road to Excellence: The Acquisition of Expert Performance in the Arts and Sciences, Sports, and Games* (Mahwah, NJ: Erlbaum, 1996); Janet Starkes and K. Anders Ericsson, eds., *Expert Performance in Sport: Recent Advances in Research on Sport Expertise* (Champaign, IL: Human Kinetics, 2003); K. Anders Ericsson, Neil Charness, Paul Feltovich, and Robert R. Hoffman, eds., *The Cambridge Handbook of Expertise and Expert Performance* (Cambridge, UK: Cambridge University Press, 2006); K. Anders Ericsson, ed., *Development of Professional Expertise: Toward Measurement of Expert Performance and*

Design of Optimal Learning Environments (Cambridge, UK: Cambridge University Press, 2009).

Chapter 1: The Power of Purposeful Practice

3 *University of Pennsylvania:* Pauline R. Martin and Samuel W. Fernberger, "Improvement in memory span," *American Journal of Psychology* 41, no. 1 (1929): 91–94.

4 *just under nine digits:* The average number of digits remembered, or "digit span," was calculated as follows: Each right answer followed by a wrong answer was assumed to be evidence that Steve had reached the limit of his digit-span memory. Thus, if he got six digits correct and followed that by getting seven digits wrong, we assumed that his digit span was somewhere between 6 and 7, and we assigned a score that was midway between the two, i.e., 6.5. At the end of the session we averaged all of the scores to get a score for the entire session. Steve's average score of 8.5 for the fourth session indicates that he could usually remember an eight-digit number and usually missed a nine-digit number, although there were plenty of exceptions because some strings were naturally easier to remember than others.

7 *not be admitted to Juilliard now:* Anthony Tommasini, "Virtuosos becoming a dime a dozen," *New York Times,* August 12, 2011, available at http://www.nytimes.com/2011/08/14/arts/music/yuja-wang-and-kirill-gerstein-lead-a-new-piano-generation.html?_r=2 (accessed November 12, 2015).

8 *type up to 212 words per minute:* http://rcranger.mysite.syr.edu/dvorak/blackburn.htm (accessed November 16, 2015).

rode 562 miles on a bicycle in twenty-four hours: http://www.guinnessworldrecords.com/world-records/greatest-distance-cycled-in-24-hours-(unpaced)-/ (accessed November 16, 2015).

calculate the roots of twelve large numbers: http://www.guinnessworldrecords.com/world-records/most-mental-calculations-in-one-minute (accessed November 16, 2015).

9 *faster than anyone else:* Personal communication (e-mail) from Bob J. Fisher, June 18, 2012.

14 *all the time:* Steve Oare, "Decisions made in the practice room: A qualitative study of middle school students' thought processes while practicing," *Update: Applications of Research in Music Education* 30 (2012): 63–70, conversation at 63.

18 *applies to physicians:* Niteesh K. Choudhry, Robert H. Fletcher, and Stephen B. Soumerai, "Systematic review: The relationship between clinical experience and quality of health care," *Annals of Internal Medicine* 142 (2005): 260–273. See also Paul M. Spengler and Lois A. Pilipis, "A comprehensive meta-analysis of the robustness of the experience-accuracy effect in clinical judgment," *Journal of Counseling Psychology* 62, no. 3 (2015): 360–378.

a consensus conference in 2015: The report from the conference can be downloaded at http://macyfoundation.org/publications/publication/enhancing-health-professions-education-technology.

Franklin's chess skills: The stories about Ben Franklin and chess are relatively well known, at least in the chess world. See, for example, John McCrary, "Chess and Benjamin Franklin—His pioneering contributions," www.benfranklin300.org/_etc_pdf/Chess_John_McCrary.pdf (accessed April 13, 2015). See also Bill Wall, "Ben Franklin and chess trivia" (2014), www.chess.com/blog/billwall/benjamin-franklin-and-chess-trivia (accessed April 13, 2015).

20 *faster than Perlman:* Christopher L. Tyner, "Violin teacher Dorothy DeLay: Step by step, she helps students reach beyond their limits," Investors.com (October 2, 2000), http://news.investors.com/management-leaders-in-success/100200-350315-violin-teacher-dorothy-delay-step-by-step-she-helps-students-reach-beyond-their-limits.htm#ixzz3D8B3Ui6D (accessed March 13, 2015).

24 *times of day:* William G. Chase and K. Anders Ericsson, "Skilled memory," in *Cognitive Skills and Their Acquisition,* ed. John R. Anderson (Hillsdale, NJ: Lawrence Erlbaum Associates, 1981), 141–189.

25 *short-term memory:* William G. Chase and K. Anders Ericsson, "Skill and working memory," in *The Psychology of Learning and Motivation,* ed. Gordon H. Bower, vol. 16 (New York: Academic Press, 1982), 1–58; K. Anders Ericsson, "Memory skill," *Canadian Journal of Psychology* 39, no. 2 (1985): 188–231; K. Anders Ericsson and Walter Kintsch, "Long-term working memory," *Psychological Review* 102 (1995): 211–245.

Chapter 2: Harnessing Adaptability

28 *difficult test in the world:* Many of the details about the testing of the London taxi-driver candidates have been taken from Jody Rosen, "The knowledge, London's legendary taxi-driver test, puts up a fight in the age of GPS," *New York Times,* December 7, 2014, http://tmagazine.blogs.nytimes.com/2014/11/10/london-taxi-test-knowledge/.

30 *were not taxi drivers:* Eleanor A. Maguire, David G. Gadian, Ingrid S. Johnsrude, Catriona D. Good, John Ashburner, Richard S. J. Frackowiak, and Christopher D. Frith, "Navigation-related structural change in the hippocampi of taxi drivers," *Proceedings of the National Academy of Sciences USA* 97 (2000): 4398–4403.

31 *food in different places:* John R. Krebs, David F. Sherry, Susal D. Healy, V. Hugh Perry, and Anthony L. Vaccarino, "Hippocampal specialization of food-storing birds," *Proceedings of the National Academy of Sciences USA* 86 (1989): 1388–1392.

food-storing experiences: Nicola S. Clayton, "Memory and the hippocampus in food-storing birds: A comparative approach," *Neuropharmacology* 37 (1998): 441–452.

than in the other subjects: In particular, the taxi drivers had more gray matter in their posterior hippocampi than those who didn't drive taxis. The gray matter is the brain tissue that contains most of the brain's neurons.

posterior hippocampi were: Strictly speaking, it was only the right posterior hippocampus that exhibited a significant increase in size with an increase in time spent as a taxi driver. Although humans have two hippocampi, for the sake of simplicity I have simply referred to the hippocampus in general. Both hippocampi are larger in London taxi drivers than in others, but the original study by Maguire et al. found a significant relationship between size and time spent driving only for the right posterior hippocampus. It is quite possible that the relationship exists on both sides but that there were too few subjects in the study for the relationship to become statistically significant.

London bus drivers: Eleanor A. Maguire, Katherine Woollett, and Hugo J. Spiers, "London taxi drivers and bus drivers: A structural MRI and neuropsychological analysis," *Hippocampus* 16 (2006): 1091–1101.

Maguire addressed this issue: Katherine Woollett and Eleanor A. Maguire, "Acquiring 'the knowledge' of London's layout drives structural brain changes," *Current Biology* 21 (2011): 2109–2114.

32 *had not ever trained at all:* Not all of the original subjects took part in the second series of measurements. All thirty-one of the control subjects returned, but only fifty-nine of the seventy-nine trainees did—thirty-nine of the forty-one who passed the tests and became licensed drivers, but only twenty of the thirty-eight who were not successful in becoming licensed cabbies.

34 *brains of blind or deaf people:* For a review, see Lofti B. Merabet and Alvaro Pascual-Leone, "Neural reorganization following sensory loss," *Nature Reviews Neuroscience* 11, no. 1 (2010): 44–52.

reroutes some of its neurons: For a clear review of what is known about neuroplasticity and blindness, see Andreja Bubic, Ella Striem-Amit, and Amir Amedi, "Large-scale brain plasticity following blindness and the use of sensory substitution devices," in *Multisensory Object Perception in the Primate Brain,* ed. Marcus Johannes Naumer and Jochen Kaiser (New York: Springer, 2010), 351–380.

35 *raised dots that make up the Braille letters:* H. Burton, A. Z. Snyder, T. E. Conduro, E. Akbudak, J. M. Ollinger, and M. E. Raichle, "Adaptive changes in early and late blind: A fMRI study of Braille reading," *Journal of Neurophysiology* 87, no. 1 (2002): 589–607. Also see Norihiro Sadato, "How the blind 'see' Braille: Lessons from functional magnetic resonance imaging," *Neuroscientist* 11, no. 6 (2005): 577–582.

fingers had been touched: Annette Sterr, Matthias M. Müller, Thomas Elbert, Brigitte Rockstroh, Christo Pantev, and Edward Taub, "Perceptual correlates of changes in cortical representation of fingers in blind multifinger Braille readers," *Journal of Neuroscience* 18, no. 11 (1998): 4417–4423.

36 *reported in 2012:* Uri Polat, Clifton Schor, Jian-Liang Tong, Ativ Zomet, Maria Lev, Oren Yehezkel, Anna Sterkin, and Dennis M. Levi, "Training the brain to overcome the effect of aging on the human eye," *Scientific Reports* 2 (2012): 278, doi:10.1038/srep00278.

39 *study on rats:* James A. Carson, Dan Nettleton, and James M. Reecy, "Differential gene expression in the rat soleus muscle during early work overload-induced hypertrophy," *FASEB Journal* 16, no. 2 (2002): 207–209.

counted 112 different genes: To be completely accurate, the researchers detected 112 mRNAs, or messenger RNAs, in the cells of the muscles that had been caused to work harder. Messenger RNAs are part of the process by which the information in DNA is used to direct the creation of proteins, and each mRNA will be associated with a particular gene, but it was the mRNAs, not the genes, that the researchers actually detected.

handle the increased workload: Again, to be completely accurate, the rats were sacrificed and their muscle tissue analyzed before their muscles could completely adjust to the new workload. This was necessary because once the muscles had adjusted and homeostasis had been regained, the muscle tissue would no longer be expressing all of these 112 genes. But if the rats had been allowed to live long enough, their muscles would have adjusted, and homeostasis would have been regained.

40 *form new brain cells:* Fred H. Gage, "Neurogenesis in the adult brain," *Journal of Neuroscience* 22 (2002): 612–613.

41 *getting rid of old ones:* Samuel J. Barnes and Gerald T. Finnerty, "Sensory experience and cortical rewiring," *Neuroscientist* 16 (2010): 186–198.

one has already learned: Arne May, "Experience-dependent structural plasticity in the adult human brain," *Trends in Cognitive Sciences* 15, no. 10 (2011): 475–482. See also Joenna Driemeyer, Janina Boyke, Christian Gaser, Christian Büchel, and Arne May, "Changes in gray matter induced by learning—Revisited," *PLoS ONE* 3 (2008): e2669.

extraordinary musical performance: An excellent review of this research can be found in Karen Chan Barrett, Richard Ashley, Dana L. Strait, and Nina Kraus, "Art and science: How musical training shapes the brain," *Frontiers in Psychology* 4, article 713 (2013). A number of the details from this section of the book come from this article and the references that it cites.

42 *1995 in the journal* Science*:* Thomas Elbert, Christo Pantev, Christian Wienbruch, Brigitte Rockstroh, and Edward Taub, "Increased cortical representation of the fingers of the left hand in string players," *Science* 270 (1995): 305–307.

responded to each touch: Because of the difficulty involved in performing magnetoencephalography, the researchers did not map out each finger on the left hand but instead looked only at the thumb and little finger. Since the brain regions corresponding to the three middle fingers would lie between the brain regions for the thumb and little finger, the researchers were able to map out the size of the region controlling the thumb and four fingers by just looking at the two digits.

43 *larger the cerebellum is:* Siobhan Hutchinson, Leslie Hui-Lin Lee, Nadine Gaab, and Gottfried Schlaug, "Cerebellar volume of musicians," *Cerebral Cortex* 13 (2003): 943–949.

guiding movements in space: Christian Gaser and Gottfried Schlaug, "Brain structures differ between musicians and non-musicians," *Journal of Neuroscience* 23 (2003): 9240–9245.

44 *than in nonmathematicians:* Kubilay Aydina, Adem Ucarb, Kader Karli Oguzc, O. Ozmen Okurd, Ayaz Agayevb, Z. Unale, Sabri Yilmazband, and Cengizhan Ozturkd, "Increased gray matter density in the parietal cortex of mathematicians: A voxel-based morphometry study," *American Journal of Neuroradiology* 28 (2007): 1859–1864.

abstract mathematical thinking: Sandra F. Witelson, Debra L. Kigar, and Thomas Harvey, "The exceptional brain of Albert Einstein," *The Lancet* 353 (1999): 2149–2153.

person was born with: Interestingly, that correlation between length of time as a mathematician and size of the region was not found for the left inferior parietal lobule. However, that may simply have been a matter of not having enough subjects in the study to be able to get a statistically valid result, and with a larger study the correlation might appear.

supplementary eye field: Tosif Ahamed, Motoaki Kawanabe, Shin Ishii, and Daniel E. Callan, "Structural differences in gray matter between glider pilots and non-pilots: A voxel-based morphometry study," *Frontiers in Neurology* 5 (2014): 248.

45 *movements of the body:* Gaoxia Wei, Yuanchao Zhang, Tianzi Jiang, and Jing Luo, "Increased cortical thickness in sports experts: A comparison of diving players with the controls," *PLoS One* 6, no. 2 (2011): e17112.

practicing in childhood: Sara L. Bengtsson, Zoltán Nagy, Stefan Skare, Lea Forsman, Hans Forssberg, and Fredrik Ullén, "Extensive piano practicing has regionally specific effects on white matter development," *Nature Neuroscience* 8 (2005): 1148–1150.

46 *provides perhaps the best example:* Katherine Woollett and Eleanor A. Maguire, "Acquiring 'the knowledge' of London's layout drives structural brain changes," *Current Biology* 21 (2011): 2109–2114.

find it difficult to walk: David Williams, Andre Kuipers, Chiaki Mukai, and Robert Thirsk, "Acclimation during space flight: Effects on human physiology," *Canadian Medical Association Journal* 180 (2009): 1317–1323.

bed for a month or so: Iñigo Mujika and Sabino Padilla, "Detraining: Loss of training-induced physiological and performance adaptations. Part II: Long-term insufficient training stimulus," *Sports Medicine* 30 (2000): 145–154.

47 *never been taxi drivers:* Katherine Woollett, Hugo J. Spiers, and Eleanor A. Maguire, "Talent in the taxi: A model system for exploring expertise," *Philosophical Transactions of the Royal Society B* 364 (2009): 1407–1416.

Chapter 3: Mental Representations

50 *best chess players in the area:* Many of the details about Alekhine and his dramatic simultaneous blindfold chess exhibition are drawn from Eliot Hearst and John Knott, *Blindfold Chess: History, Psychology, Techniques, Champions, World Records, and Important Games* (Jefferson, NC: McFarland, 2009).

51 *no literal blindfold involved:* Details on the history of blindfold chess can be found in many places, but the most comprehensive is Hearst and Knott, ibid.

two losses, and nineteen draws: Eliot Hearst, "After 64 years: New world blindfold record set by Marc Lang playing 46 games at once," Blindfold Chess, December 16, 2011, http://www.blindfoldchess.net/blog/2011/12/after_64_years_new_world_blindfold_record_set_by_marc_lang_playing_46_games/ (accessed May 27, 2015).

52 *when he was seven:* Details on Alekhine's life and chess-playing career are from several sources: Alexander Kotov, *Alexander Alekhine,* trans. K. P. Neat (Albertson, NY: R. H. M. Press, 1975); Hearst and Knott, *Blindfold Chess;* "Alekhine's biography" on Chess.com, www.chess.com/groups/forumview/alekhines-biography2 (accessed May 27, 2015); and "Alexander Alekhine" on Chessgames.com, www.chessgames.com/perl/chessplayer?pid=10240 (accessed May 27, 2015).

"White wins!": Kotov, *Alexander Alekhine.*

correspondence tournaments: Hearst and Knott, *Blindfold Chess,* 74.

54 *of course, to himself:* Alexander Alekhine, *On the Road to a World Championship, 1923–1927,* 1st English ed. (New York: Pergamon Press, 1984), as quoted in Hearst and Knott, *Blindfold Chess,* 78.

55 *board almost perfectly:* Adrianus D. De Groot, *Thought and Choice in Chess,* 2nd ed. (The Hague: Mouton de Gruyter, 1978).

simple but effective experiment: William G. Chase and Herbert A. Simon, "Perception in chess," *Cognitive Psychology* 4 (1973): 55–81. The experiment comparing the memory of a master and a novice on normal chess positions and random collections of chess pieces was actually first carried out by Adriaan de Groot. See, for example, Adrianus Dingeman De Groot, *Thought and Choice in Chess* (The Hague: Mouton, 1965) and Adrianus Dingeman De Groot, "Perception and memory versus thought: Some old ideas and recent findings," in *Problem Solving,* ed. B. Kleimnuntz (New York: Wiley, 1966), 19–50.

56 *reiterated the original findings:* Fernand Gobet and Neil Charness, "Expertise in chess," in *The Cambridge Handbook of Expertise and Expert Performance,* ed. K. Anders Ericsson, Neil Charness, Paul J. Feltovich, and Robert R. Hoffman (New York: Cambridge University Press, 2006), 523–538.
shown with verbal memory: William G. Chase and K. Anders Ericsson, "Skill and working memory," in *The Psychology of Learning and Motivation,* ed. G. H. Bower (New York: Academic Press, 1982), 1–58.

57 *fifty thousand of these chunks:* Herbert A. Simon and Kevin Gilmartin, "A simulation of memory for chess positions," *Cognitive Psychology* 5, no. 1 (1973): 29–46.
higher-level patterns: Hartmut Freyhof, Hans Gruber, and Albert Ziegler, "Expertise and hierarchical knowledge representation in chess," *Psychological Research* 54 (1992): 32–37.
"lines of force" and "power": See, for example, Hearst and Knott, *Blindfold Chess,* 10.

60 *visuospatial abilities:* Andrew Waters, Fernand Gobet, and Gery Leyden, "Visuo-spatial abilities in chess players," *British Journal of Psychology* 93 (2002): 557–565.

62 *reflexes are no faster:* Sean Müller and Bruce Abernethy, "Expert anticipatory skill in striking sports: A review and a model," *Research Quarterly for Exercise and Sport* 83, no. 2 (2012): 175–187.

64 *happened on the field:* Paul Ward, K. Anders Ericsson, and A. Mark Williams, "Complex perceptual-cognitive expertise in a simulated task environment," *Journal of Cognitive Engineering and Decision Making* 7 (2013): 231–254.

65 *indoor rock climbing:* Bettina E. Bläsing, Iris Güldenpenning, Dirk Koester, and Thomas Schack, "Expertise affects representation structure and categorical activation of grasp postures in climbing," *Frontiers in Psychology* 5 (2014): 1008.

67 *understands about the sport:* For a general review and list of references on the subject of reading comprehension and mental representations, see K. Anders Ericsson and Walter Kintsch, "Long-term working memory," *Psychological Review* 102, no. 2 (1995): 211–245.

69 *two hundred who sent in answers:* Lisa Sanders, "Think like a doctor: A knife in the ear," *New York Times,* March 23, 2011, http://well.blogs.nytimes.com/2015/08/06/think-like-a-doctor-a-knife-in-the-ear/ (accessed September 24, 2015); Lisa Sanders, "Think like a doctor: A knife in the ear solved," *New York Times,* March 24, 2011, http://well.blogs.nytimes.com/2015/08/07/think-like-a-doctor-a-knife-in-the-ear-solved/ (accessed September 24, 2015).

71 *the most likely one:* Vimla L. Patel, Jose F. Arocha, and David R. Kaufmann, "Diagnostic reasoning and medical expertise," in *The Psychology of Learning and Motivation,* ed. Douglas Medin, vol. 30 (New York: Academic Press, 1994), 187–251.

72 *in 150 agents:* Thomas W. Leigh, Thomas E. DeCarlo, David Allbright, and James Lollar, "Salesperson knowledge distinctions and sales performance," *Journal of Personal Selling & Sales Management* 34, no. 2 (2014): 123–140.
from hold to hold: Xavier Sanchez, P. Lambert, G. Jones, and D. J. Llewellyn, "Efficacy of pre-ascent climbing route visual inspection in indoor sport climbing," *Scandinavian Journal of Medicine & Science in Sports* 22, no. 1 (2010): 67–72.
making the first incision: See, for example, Nathan R. Zilbert, Laurent St-Martin, Glenn Regehr, Steven Gallinger, and Carol-Anne Moulton, "Planning to avoid trouble in the operating room: Experts' formulation of the preoperative plan," *Journal of Surgical Education* 72, no. 2 (2014): 271–277.

73 *used in writing an essay:* As told in Marlene Scardamalia and Carl Bereiter, "Knowledge telling and knowledge transforming in written composition," in *Advances in Applied Psycholinguistics,* ed. Sheldon Rosenberg (Cambridge, UK: Cambridge University Press, 1987), 142–175. See page 149 in particular.

74 *"knowledge telling":* The terms "knowledge telling" and "knowledge transforming" come from Scardamalia and Bereiter, ibid.

77 *representations the best ones create:* For a good overview, see Paul L. Sikes, "The effects of specific practice strategy use on university string players' performance," *Journal of Research in Music Education* 61, no. 3 (2013): 318–333.
more or less effective: Gary E. McPherson and James M. Renwick, "A longitudinal study of self-regulation in children's music practice," *Music Education Research* 3, no. 2 (2001): 169–186.

79 *three thousand music students:* Susan Hallam, Tiija Rinta, Maria Varvarigou, Andrea Creech, Ioulia Papageorgi, Teresa Gomes, and Jennifer Lanipekun, "The development of practicing strategies in young people," *Psychology of Music* 40, no. 5 (2012): 652–680.

80 *performs a piece of music:* Roger Chaffin and Gabriela Imreh, "'Pulling teeth and torture': Musical memory and problem solving," *Thinking and Reasoning* 3, no. 4 (1997): 315–336; Roger Chaffin and Gabriela Imreh, "A comparison of practice and self-report as sources of information about the goals of expert practice," *Psychology of Music* 29 (2001): 39–69; Roger Chaffin, Gabriela Imreh, Anthony F. Lemieux, and Colleen Chen, "'Seeing the big picture': Piano playing as expert problem solving," *Music Perception* 20, no. 4 (2003): 465–490.

81 *learning to play a piece:* Roger Chaffin and Topher Logan, "Practicing perfection: How concert soloists prepare for performance," *Advances in Cognitive Psychology* 2, nos. 2–3 (2006): 113–130.

Chapter 4: The Gold Standard

84 *World Memory Sports Council:* The memory competition statistics as of July 2015 are from the website of the World Memory Sports Council, http://www

.world-memory-statistics.com/discipline.php?id=spoken1 (accessed July 15, 2015).

86 *without writing them down:* K. Anders Ericsson and Peter G. Polson, "A cognitive analysis of exceptional memory for restaurant orders," in *The Nature of Expertise,* ed. Michelene T. H. Chi, Robert Glaser, and Marshall J. Farr (Hillsdale, NJ: Lawrence Erlbaum, 1988), 23–70.

they began a new play: William L. Oliver and K. Anders Ericsson, "Repertory actors' memory for their parts," in *Eighth Annual Conference of the Cognitive Society* (Hillsdale, NJ: Lawrence Erlbaum Associates, 1986), 399–406.

87 *musical accomplishment:* Some details were provided in an earlier publication: K. Anders Ericsson, Clemens Tesch-Römer, and Ralf Krampe, "The role of practice and motivation in the acquisition of expert-level performance in real life: An empirical evaluation of a theoretical frame-work," in *Encouraging the Development of Exceptional Skills and Talents,* ed. Michael J. A. Howe (Leicester, UK: British Psychological Society, 1990), 109–130. However, the complete account of the study was provided in K. Anders Ericsson, Clemens Tesch-Römer, and Ralf Krampe, "The role of deliberate practice in the acquisition of expert performance," *Psychological Review* 100, no. 3 (1993): 363–406.

88 *good, better, and best:* We did not rely simply on the judgments of the faculty. We verified those judgments with other measures. In particular, we collected information on how well the students had done in open music competitions, and we found that our "best" violinists had had greater success than the "better" violinists and that both of those groups had had more success than the music-education students. We also found that the violinists in the best group could play significantly more music from memory than the violinists in the better group, and violinists in both groups knew more music by memory than the future music teachers. So we were comfortable that we really had assembled three groups of violinists with clearly distinct abilities.

92 *their practice histories:* Although we had to rely on years-old memories of how much they had practiced earlier in their lives, we believed that their memories were likely to be reasonably accurate. From almost the very beginning, these violinists were setting aside a certain amount of time each day or each week to practice — an amount of time that steadily increased as they got older — and so they were very aware of how much practice time they were putting in at each stage.

93 *relatively accurate:* One possible issue was that the different groups of students might have had different biases in their estimates of how much they practiced. However, if there had been such biases, one would expect that the very best students — who had heard all their lives how talented they were — would have bought into the idea that they did not need to practice as much as other, less-talented students and had thus consistently underestimated how much time they had spent practicing. Thus, any bias should have

made it less likely that we would see an effect with the better students having practiced more.

95 *played in their achievements:* Carla U. Hutchinson, Natalie J. Sachs-Ericsson, and K. Anders Ericsson, "Generalizable aspects of the development of expertise in ballet across countries and cultures: A perspective from the expert performance approach," *High Ability Studies* 24 (2013): 21–47.

96 *decade of intense study:* Herbert A. Simon and William G. Chase, "Skill in chess," *American Scientist* 61 (1973): 394–403.

97 *to become a grandmaster:* The trend to younger grandmasters: Robert W. Howard, "Preliminary real-world evidence that average human intelligence really is rising," *Intelligence* 27, no. 3 (1999): 235–250. Evidence for more effective training methods: Fernand Gobet, Guillermo Campitelli, and Andrew J. Waters, "Rise of human intelligence: Comments on Howard" (1999), *Intelligence* 30, no. 4 (2002): 303–311.

98 *"deliberate practice":* Ericsson, Tesch-Römer, and Krampe, "The role of deliberate practice," 367–368.

101 *remember more than fifteen:* David Wechsler, *The Range of Human Capacities* (New York: Williams & Wilkins, 1935).

Feng Wang of China: K. Anders Ericsson, Xiaojun Cheng, Yafeng Pan, Yixuan Ku, and Yi Hu. "Refined memory encodings mediate exceptional memory span in a world-class memorist" (paper submitted for publication), corresponding author Yi Hu, School of Psychology and Cognitive Science, East China Normal University, Shanghai, China.

102 *remember large amounts of information:* Frances A. Yates, *The Art of Memory* (Chicago: University of Chicago Press, 1966).

imposed by short-term memory: For a more detailed discussion of the use of long-term memory in the way, see K. Anders Ericsson and W. Kintsch, "Long-term working memory," *Psychological Review* 102 (1995): 211–245.

104 *physical attractiveness:* Alf Gabrielsson, "The performance of music," in *The Psychology of Music,* ed. Diana Deutsch, 2nd ed. (San Diego, CA: Academic Press, 1999), 501–602.

suggested an experiment: Robert T. Hodgson, "An examination of judge reliability at a major U.S. wine competition," *Journal of Wine Economics* 3, no. 2 (2008): 105–113.

105 *received minimal training:* Robyn M. Dawes, *House of Cards: Psychology and Psychotherapy Built on Myth* (New York: Free Press, 1994).

novices or random chance: One of the earliest studies was Carl-Axel S. Staël Von Holstein, "Probabilistic forecasting: An experiment related to the stock market," *Organizational Behavior and Human Performance* 8, no. 1 (1972): 139–158. Staël Von Holstein studied the stock-price predictions of stock-market experts, bankers, statisticians, university business professors, and university business teachers over a twenty-week period and found that, on average, none of the groups performed significantly better than would be expected by random chance. For a more recent review, see K. Anders Ericsson, Patric Andersson, and Edward T. Cokely, "The enigma of financial expertise: Superior and reproducible investment performance in efficient markets," http://citeseerx.ist.psu.edu/viewdoc/download?, doi:10.1.1.337.3918&rep=rep1&type=pdf (accessed August 16, 2015).

types of doctors and nurses: K. Anders Ericsson, "Acquisition and maintenance of medical expertise: A perspective from the expert-performance approach with deliberate practice," *Academic Medicine* 90 (2015): 1471–1486. See also Niteesh K. Choudhry, Robert H. Fletcher, and Stephen B. Soumerai, "Systematic review: The relationship between clinical experience and quality of health care," *Annals of Internal Medicine* 142 (2005): 260–273; K. Anders Ericsson, James Whyte 4th, and Paul Ward, "Expert performance in nursing: Reviewing research on expertise in nursing within the framework of the expert performance approach," *Advances in Nursing Science* 30, no. 1 (2007): E58–E71; Paul M. Spengler, Michael J. White, Stefanía Ægisdóttir, Alan S. Maugherman, Linda A. Anderson, Robert S. Cook, Cassandra N. Nichols, Georgios K. Lampropoulos, Blain S. Walker, Genna R. Cohen, and Jeffrey D. Rush, "The meta-analysis of clinical judgment project: Effects of experience on judgment accuracy," *Counseling Psychology* 20 (2009): 350–399.

107 *able to use the same methods:* Those methods are described in K. Anders Ericsson, "Protocol analysis and expert thought: Concurrent verbalizations of thinking during experts' performance on representative task," in *The Cambridge Handbook of Expertise and Expert Performance,* ed. K. Anders Ericsson, Neil Charness, Paul Feltovich, and Robert R. Hoffman (Cambridge, UK: Cambridge University Press, 2006), 223–242.

109 *Malcolm Gladwell's* Outliers*:* Malcolm Gladwell, *Outliers: The Story of Success* (New York: Little, Brown), 2008.

111 *biography of the Beatles by Mark Lewisohn:* Mark Lewisohn, *Tune In* (New York: Crown Archetype, 2013).

students or the ballet dancers: Even some researchers forget this from time to time. As I was working on this book, a group of researchers published a meta-analysis — that is, an analysis of a large number of previously published studies — that concluded that structured practice (although they called it "deliberate practice") explained relatively little of the difference in performance among individuals in various fields, including music, sports, education, and other professions. See Brooke N. Macnamara, David Z. Hambrick, and Frederick L. Oswald, "Deliberate practice and performance in music, games, sports, education, and professions: A meta-analysis," *Psychological Science* 25 (2014): 1608–1618. The major problem with this meta-analysis was that few of the studies the researchers examined were actually looking at the effects of the type of practice on performance that we had referred to as deliberate practice; instead, the researchers used very loose criteria to decide which studies to include in their meta-analysis, so they ended up examining a collection of studies that dealt mainly with various types of practice and training that did not meet the criteria of deliberate practice as we described it earlier in this chapter. I offer a detailed critique of their work in K. Anders Ericsson, "Challenges for the estimation of an upper-bound on relations between accumulated deliberate practice and the associated performance in domains of expertise: Comments on Macnamara, Hambrick, and Oswald's (2014) published meta-analysis," available on my website, https://psy.fsu.edu/faculty/ericsson/ericsson.hp.html. The bottom line is that what their meta-analysis really demonstrated is that if you wish to understand why some people perform better than others, it is not sufficient to attempt to measure all hours engaged in just any sort of practice; you need to focus on the activities based on our criteria for deliberate practice. See, for instance, the discussion in K. Anders Ericsson, "Why expert performance is special and cannot be extrapolated from studies of performance in the general population: A response to criticisms," *Intelligence* 45 (2014): 81–103.

improve particular aspects of performance: See, for example, the definition of deliberate practice found in K. Anders Ericsson and Andreas C. Lehmann, "Expert and exceptional performance: Evidence of maximal adaptations to task constraints," *Annual Review of Psychology* 47 (1996): 273–305. Deliberate practice consists of "individualized training activities specially designed by a coach or teacher to improve specific aspects of an individual's performance through repetition and successive refinement" (278–279).

112 *first published research:* Ericsson, Tesch-Römer, and Krampe, "The role of deliberate practice."

never less than ten years: John R. Hayes, *The Complete Problem Solver* (Philadelphia: Franklin Institute Press, 1981).

113 *"ten thousand hours is a mental disorder":* Scott Adams, *Dilbert,* February 7, 2013.

Chapter 5: Principles of Deliberate Practice on the Job

115 *It was 1968:* Details about the creation and early days of the Top Gun school are taken from Ralph Earnest Chatham, "The 20th-century revolution in military training," in *Development of Professional Expertise,* ed. K. Anders Ericsson (New York: Cambridge University Press, 2009), 27–60. See also Robert K. Wilcox, *Scream of Eagles* (New York: Pocket Star Books, 1990).

117 *the real action occurred:* Chatham, "The 20th-century revolution."

The results of this training were dramatic: "'You fight like you train,' and Top Gun crews train hard," *Armed Forces Journal International* 111 (May 1974): 25–26, 34.

118 *the most dominant performance:* Wilcox, *Scream of Eagles,* vi.

119 *navy did it mainly through trial and error:* Ibid.

120 *most resemble deliberate practice:* K. Anders Ericsson, "The influence of experience and deliberate practice on the development of superior expert performance," in *Cambridge Handbook of Expertise and Expert Performance,* ed. K. Anders Ericsson, Neil Charness, Paul Feltovich, and Robert R. Hoffman (Cambridge, UK: Cambridge University Press, 2006), 685–706.

an article in Fortune *magazine:* Geoff Colvin, "What it takes to be great: Research now shows that the lack of natural talent is irrelevant to great success. The secret? Painful and demanding practice and hard work," *Fortune,* October 19, 2006, http://archive.fortune.com/magazines/fortune/fortune_archive/2006/10/30/8391794/index.htm (accessed September 27, 2015).

121 *Art has fully embraced the deliberate-practice mindset:* Many of the details I offer here can be found on Turock's website, www.turock.com, and in a book he wrote: Art Turock, *Competent Is Not an Option: Build an Elite Leadership Team Following the Talent Development Game Plan of Sports Champions* (Kirkland, WA: Pro Practice Publishing, 2015).

123 *Blue Bunny ice cream company:* Turock tells the Blue Bunny story in his book *Competent Is Not an Option,* ibid.

125 *perform this job much better than others:* Diana L. Miglioretti, Charlotte C. Gard, Patricia A. Carney, Tracy L. Onega, Diana S. M. Buist, Edward A. Sickles, Karla Kerlikowske, Robert D. Rosenberg, Bonnie C. Yankaskas, Berta M. Geller, and Joann G. Elmore, "When radiologists perform best: The learning curve in screening mammogram interpretation," *Radiology* 253 (2009): 632–640. See also Calvin F. Nodine, Harold L. Kundel, Claudia Mello-Thoms, Susan P. Weinstein, Susan G. Orel, Daniel C. Sullivan, and Emily F. Conant, "How experience and training influence mammography expertise," *Academic Radiology* 6 (1999): 575–585.

A 2004 analysis: William E. Barlow, Chen Chi, Patricia A. Carney, Stephen H. Taplin, Carl D'Orsi, Gary Cutter, R. Edward Hendrick, and Joann G. Elmore, "Accuracy of screening mammography interpretation by characteristics of radiologists," *Journal of the National Cancer Institute* 96 (2004): 1840–1850.

126 *regularly request unnecessary biopsies:* Ibid.

meeting of the American Association of Medical Colleges: K. Anders Ericsson, "Deliberate practice and the acquisition and maintenance of expert performance in medicine and related domains," *Academic Medicine* 79 (2004): S70–S81.

127 *similar to what I proposed:* See http://www.breastaustralia.com/public/index.

interpreted mammograms in their professional practice: BaoLin Pauline Soh, Warwick Bruce Lee, Claudia Mello-Thoms, Kriscia Tapia, John Ryan, Wai Tak Hung, Graham Thompson, Rob Heard, and Patrick Brennan, "Certain performance values arising from mammographic test set readings correlate well with clinical audit," *Journal of Medical Imaging and Radiation Oncology* 59 (2015): 403–410.

a set of 234 cases: M. Pusic, M. Pecaric, and K. Boutis, "How much practice is enough? Using learning curves to assess the deliberate practice of radiograph interpretation," *Academic Medicine* 86 (2011): 731–736.

128 *the best radiologists have indeed developed:* Alan Lesgold, Harriet Rubinson, Paul Feltovich, Robert Glaser, Dale Klopfer, and Yen Wang, "Expertise in a complex skill: Diagnosing X-ray pictures," in *The Nature of Expertise,* ed. Michelene T. H. Chi, Robert Glaser, and Marshall J. Farr (Hillsdale, NJ: Lawrence Erlbaum Associates, 1988), 311–342; Roger Azevedo, Sonia Faremo, and Susanne P. Lajoie, "Expert-novice differences in mammogram interpretation," in *Proceedings of the 29th Annual Cognitive Science Society,* ed. D. S. McNamara and J. G. Trafton (Nashville, TN: Cognitive Science Society, 2007), 65–70.

the types of cases and lesions: Claudia Mello-Thoms, Phuong Dung Trieu, and Mohammed A. Rawashdeh, "Understanding the role of correct lesion assessment in radiologists' reporting of breast cancer," in *Breast Imaging: Proceedings, 12th International Workshop, IWDM 2014,* ed. Hiroshi Fujita, Takeshi Hara, and Chisako Muramatsu (Cham, Switzerland: Springer International, 2014), 341–347.

129 *In almost every case, these injuries were due to:* Lawrence L. Way, L. Stewart, W. Gantert, Kingsway Liu, Crystine M. Lee, Karen Whang, and John G. Hunter, "Causes and prevention of laparoscopic bile duct injuries: Analysis of 252 cases from a human factors and cognitive psychology perspective," *Annals of Surgery* 237, no. 4 (2003): 460–469.

expert surgeons develop ways: Helena M. Mentis, Amine Chellali, and Steven Schwaitzberg, "Learning to see the body: Supporting instructional practices in laparoscopic surgical procedures," in *Proceedings of the SIGCHI Conference on Human Factors in Computing Systems* (New York: Association for Computing Machinery, 2014), 2113–2122.

130 *prior to a blood transfusion:* The blood transfusion example comes from David Liu, Tobias Grundgeiger, Penelope M. Sanderson, Simon A. Jenkins, and Terrence A. Leane, "Interruptions and blood transfusion checks: Lessons from the simulated operating room," *Anesthesia & Analgesia* 108 (2009): 219–222.

132 *an extensive review of research:* Niteesh K. Choudhry, Robert H. Fletcher, and Stephen B. Soumerai, "Systematic review: The relationship between clinical experience and quality of health care," *Annals of Internal Medicine* 142 (2005): 260–273. See also Paul M. Spengler and Lois A. Pilipis, "A comprehensive meta-analysis of the robustness of the experience-accuracy effect in clinical judgment," *Journal of Counseling Psychology* 62, no. 3 (2015): 360–378.

133 *Another study of decision-making accuracy:* Paul M. Spengler, Michael J. White, Stefanía Ægisdóttir, Alan S. Maugherman, Linda A. Anderson, Robert S. Cook, Cassandra N. Nichols, Georgios K. Lampropoulos, Blain S. Walker, Genna R. Cohen, and Jeffrey D. Rush, "The meta-analysis of clinical judgment project: Effects of experience on judgment accuracy," *Counseling Psychology* 20 (2009): 350–399.
experienced nurses do not: K. Anders Ericsson, James Whyte 4th, and Paul Ward, "Expert performance in nursing: Reviewing research on expertise in nursing within the framework of the expert performance approach," *Advances in Nursing Science* 30, no. 1 (2007): E58–E71.

134 *Davis and a group of colleagues examined:* Dave Davis, Mary Ann Thomson O'Brien, Nick Freemantle, Fredric M. Wolf, Paul Mazmanian, and Anne Taylor-Vaisey, "Impact of formal continuing medical education: Do conferences, workshops, rounds, and other traditional continuing education activities change physician behavior or health care outcomes?" *JAMA* 282, no. 9 (1999): 867–874.

135 *Forsetlund updated Davis's work:* Louise Forsetlund, Arild Bjørndal, Arash Rashidian, Gro Jamtvedt, Mary Ann O'Brien, Fredric M. Wolf, Dave Davis, Jan Odgaard-Jensen, and Andrew D. Oxman, "Continuing education meetings and workshops: Effects on professional practice and health care outcomes," *Cochrane Database of Systematic Reviews* 2 (2012): CD003030.

136 *"See one, do one, teach one":* J. M. Rodriguez-Paz, M. Kennedy, E. Salas, A. W. Wu, J. B. Sexton, E. A. Hunt, and P. J. Pronovost, "Beyond 'see one, do one, teach one': Toward a different training paradigm," *Quality and Safety in Health Care* 18 (2009): 63–68. See also William C. McGaghie, Jacob R. Suker, S. Barry Issenberg, Elaine R. Cohen, Jeffrey H. Barsuk, and Diane B. Wayne, "Does simulation-based medical education with deliberate practice yield better results than traditional clinical education? A meta-analytic comparative review of the evidence," *Academic Medicine* 86, no. 6 (June 2011): 706–711.
they found no difference: Michael J. Moore and Charles L. Bennett, and the Southern Surgeons Club, "The learning curve for laparoscopic cholecystectomy," *American Journal of Surgery* 170 (1995): 55–59.

138 *In one very relevant study:* John D. Birkmeyer, Jonathan F. Finks, Amanda O'Reilly, Mary Oerline, Arthur M. Carlin, Andre R. Nunn, Justin Dimick, Mousumi Banerjee, and Nancy J. O. Birkmeyer, "Surgical skill and complication rates after bariatric surgery," *New England Journal of Medicine* 369 (2013): 1434–1442.

139 *how we might identify expert doctors:* K. Anders Ericsson, "Acquisition and maintenance of medical expertise: A perspective from the expert performance approach and deliberate practice," *Academic Medicine* 90, no. 11 (2015): 1471–1486.
researchers led by Andrew Vickers: Andrew J. Vickers, Fernando J. Bianco,

Angel M. Serio, James A. Eastham, Deborah Schrag, Eric A. Klein, Alwyn M. Reuther, Michael W. Kattan, J. Edson Pontes, and Peter T. Scardino, "The surgical learning curve for prostate cancer control after radical prostatectomy," *Journal of the National Cancer Institute* 99, no. 15 (2007): 1171–1177.

In a follow-up study: Andrew J. Vickers, Fernando J. Bianco, Mithat Gonen, Angel M. Cronin, James A. Eastham, Deborah Schrag, Eric A. Klein, Alwyn M. Reuther, Michael W. Kattan, J. Edson Pontes, and Peter T. Scardino, "Effects of pathologic stage on the learning curve for radical prostatectomy: Evidence that recurrence in organ-confined cancer is largely related to inadequate surgical technique," *European Urology* 53, no. 5 (2008): 960–966.

140 *get better as they gain experience:* K. Anders Ericsson, "Surgical expertise: A perspective from the expert-performance approach," in *Surgical Education in Theoretical Perspective: Enhancing Learning, Teaching, Practice, and Research,* ed. Heather Fry and Roger Kneebone (Berlin: Springer, 2011), 107–121.

141 *a study of radiologists interpreting mammograms:* Diana L. Miglioretti, Charlotte C. Gard, Patricia A. Carney, Tracy L. Onega, Diana S. M. Buist, Edward A. Sickles, Karla Kerlikowske, Robert D. Rosenberg, Bonnie C. Yankaskas, Berta M. Geller, and Joann G. Elmore, "When radiologists perform best: The learning curve in screening mammogram interpretation," *Radiology* 253 (2009): 632–640.

a recent study of eight surgeons: Curtis Craig, Martina I. Klein, John Griswold, Krishnanath Gaitonde, Thomas McGill, and Ari Halldorsson, "Using cognitive task analysis to identify critical decisions in the laparoscopic environment," *Human Factors* 54, no. 3 (2012): 1–25.

142 *"even expert surgeons":* Ibid.

Think Like a Commander Training Program: James W. Lussier, Scott B. Shadrick, and Michael Prevou, *Think Like a Commander Prototype: Instructor's Guide to Adaptive Thinking* (Fort Knox, KY: Armored Forces Research Unit, U.S. Army Research Institute, 2003).

143 *recent studies done by medical researchers in Canada:* Sayra M. Cristancho, Tavis Apramian, Meredith Vanstone, Lorelei Lingard, Michael Ott, and Richard J. Novick, "Understanding clinical uncertainty: What is going on when experienced surgeons are not sure what to do?" *Academic Medicine* 88 (2013): 1516–1521; and Sayra M. Cristancho, Meredith Vanstone, Lorelei Lingard, Marie-Eve LeBel, and Michael Ott, "When surgeons face intraoperative challenges: A naturalistic model of surgical decision making," *American Journal of Surgery* 205 (2013): 156–162.

144 *stop in the middle and quiz people:* Mica R. Endsley, "Expertise and situation awareness," in *The Cambridge Handbook of Expertise and Expert Performance,* ed. K. Anders Ericsson, Neil Charness, Paul J. Feltovich, and Robert R. Hoffman, eds. (Cambridge, UK: Cambridge University Press,

2006), 633–652. See also Paul M. Salmon, Neville A. Stanton, Guy H. Walker, Daniel Jenkins, Darsha Ladva, Laura Rafferty, and Mark Young, "Measuring situation awareness in complex systems: Comparison of measures study," *International Journal of Industrial Ergonomics* 39 (2009): 490–500.

Chapter 6: Principles of Deliberate Practice in Everyday Life

145 *research in various places:* Dan McLaughlin has mentioned specifically reading about my research in *Talent Is Overrated,* but there were already at that time several books that discussed the power of deliberate practice, so the idea was well known. The books include Geoff Colvin, *Talent Is Overrated: What Really Separates World-Class Performers from Everybody Else* (New York: Portfolio, 2008); Malcolm Gladwell, *Outliers: The Story of Success* (New York: Little, Brown, 2008); and Daniel Coyle, *The Talent Code: Greatness Isn't Born. It's Grown. Here's How* (New York: Bantam Dell, 2009).

his efforts to become a professional golfer: Dan McLaughlin has a website at which he describes his plan and his progress, thedanplan.com. There is also a good story about Dan McLaughlin that was published in *Golf:* Rick Lipsey, "Dan McLaughlin thinks 10,000 hours of focused practice will get him on Tour," *Golf,* December 9, 2011, www.golf.com/tour-and-news/dan-mclaughlin-thinks-10000-hours-focused-practice-will-get-him-tour (accessed August 26, 2015).

146 *allow him to compete in PGA tournaments:* Since Dan started his plan, the rules for getting a PGA Tour card have changed. Now doing well enough at the PGA Tour Qualifying Tournament will only get you onto the PGA's Web.com tour, and you have to do well enough on that tour to get onto the PGA Tour.

he gave an interview: Lipsey, "Dan McLaughlin thinks 10,000 hours."

150 *ready to find a coach at the next level:* Personal communication from Dan McLaughlin, June 4, 2014.

151 *varying the point on the dartboards:* Linda J. Duffy, Bachman Baluch, and K. Anders Ericsson, "Dart performance as a function of facets of practice amongst professional and amateur men and women players," *International Journal of Sports Psychology* 35 (2004): 232–245.

If you want to get better at bowling: Kevin R. Harris, "Deliberate practice, mental representations, and skilled performance in bowling" (Ph.D. diss., Florida State University, 2008), Electronic Theses, Treatises and Dissertations, DigiNole Commons, paper no. 4245.

a group of Swedish researchers: Christina Grape, Maria Sandgren, Lars-Olof Hansson, Mats Ericson, and Tores Theorell, "Does singing promote well-being? An empirical study of professional and amateur singers during a singing lesson," *Integrative Physiological and Behavioral Science* 38 (2003): 65–74.

152 *high school golfers developing this sort of focus:* Cole G. Armstrong, "The influence of sport specific social organizations on the development of identity: A case study of professional golf management" (Ph.D. diss., Florida State University, 2015), Electronic Theses, Treatises and Dissertations, DigiNole Commons, paper no. 9540.

Cole quoted one high school golfer: Ibid., 179.

Natalie Coughlin once described: The details about Natalie Coughlin's training are taken from Gina Kolata, "Training insights from star athletes," *New York Times,* January 14, 2013.

153 *an extended study of Olympic swimmers:* Daniel F. Chambliss, *Champions: The Making of Olympic Swimmers* (New York: Morrow, 1988); Daniel F. Chambliss, "The mundanity of excellence: An ethnographic report on stratification and Olympic swimmers," *Sociological Theory* 7 (1989): 70–86.

"every detail becomes a firmly ingrained habit": Chambliss, "Mundanity of excellence," 85.

Researchers who have studied long-distance runners: The pioneering study was W. P. Morgan and M. L. Pollock, "Psychological characterization of the elite distance runner," *Annals of the New York Academy of Sciences* 301 (1977): 382–403. A more recent review of the subsequent research and a description of more concurrent reports of thinking can be found in Ashley Samson, Duncan Simpson, Cindra Kamphoff, and Adrienne Langlier, "Think aloud: An examination of distance runners' thought processes," *International Journal of Sport and Exercise Psychology,* online publication July 25, 2015, doi :10.1080/1612197X.2015.1069877.

155 *Early in his autobiography:* Benjamin Franklin, *The Autobiography of Benjamin Franklin* (New York: Henry Holt, 1916), original publication in French in 1791; first English printing, 1793, https://www.gutenberg.org/files/20203/20203-h/20203-h.htm (accessed August 30, 2015). I first described the method Franklin used to improve his writing in my introductory chapter in K. Anders Ericsson, ed., *Roads to Excellence: The Acquisition of Expert Performance in the Arts and Sciences, Sports, and Games* (Mahwah, NJ: Erlbaum, 1996), 1–50. A nice recent description is given by Shane Snow, "Ben Franklin taught himself to write with a few clever tricks," *The Freelancer,* August 21, 2014, http://contently.net/2014/08/21/stories/ben-franklin-taught-write-clever-tricks/ (accessed August 30, 2015).

160 *a way very similar to the technique Franklin used:* Lecoq de Boisbaudran, *The Training of the Memory in Art and the Education of the Artist,* trans. L. D. Luard (London: MacMillan, 1911), https://books.google.com/books?hl=en&lr=&id=SJufAAAAMAAJ&oi=fnd&pg=PR5&dq=the+training

+of+the+memory+in+art+and+the+education+of+the+artist&ots=CvAENj-mHl&sig=Iu4ku1d5F-uIP_aacBLugvYAiTU#v=onepage&q=the%20training%20of%20the%20memory%20in%20art%20and%20the%20education%20of%20the%20artist&f=false (accessed October 2, 2015).

163 *well-established method to get past such a plateau:* K. Anders Ericsson, "The acquisition of expert performance as problem solving," in *The Psychology of Problem Solving,* ed. Janet E. Davidson and Robert J. Sternberg (New York: Cambridge University Press, 2003), 31–83.

165 *what set apart the very best spellers:* Angela L. Duckworth, Teri A. Kirby, Eli Tsukayama, Heather Berstein, and K. Anders Ericsson, "Deliberate practice spells success: Why grittier competitors triumph at the National Spelling Bee," *Social Psychology and Personality Science* 2 (2011): 174–181.

169 *The ones who are successful in losing weight:* See, for example, Rena R. Wing and Suzanne Phelan, "Long-term weight-loss maintenance," *American Journal of Clinical Nutrition* 82 (supplement, 2005): 222S–225S; K. Ball and D. Crawford, "An investigation of psychological, social, and environmental correlates of obesity and weight gain in young women," *International Journal of Obesity* 30 (2006): 1240–1249.

172 *One of Sweden's most famous athletes:* This episode is described in Hägg's autobiography written some forty years later: Gunder Hägg, *Mitt Livs Lopp* [The competition of my life] (Stockholm: Norstedts, 1987).

175 *recruited eleven of the most intellectually interesting people:* Franklin, *Autobiography.*

Chapter 7: The Road to Extraordinary

180 *a grand experiment:* The details of the Polgár story come from a number of places: Linnet Myers, "Trained to be a genius, girl, 16, wallops chess champ Spassky for $110,000," *Chicago Tribune,* February 18, 1993, http://articles.chicagotribune.com/1993-02-18/news/9303181339_1_judit-polgar-boris-spassky-world-chess-champion (accessed August 19, 2015); Austin Allen, "Chess grandmastery: Nature, gender, and the genius of Judit Polgár," *JSTOR Daily,* October 22, 2014, http://daily.jstor.org/chess-grandmastery-nature-gender-genius-judit-polgar/ (accessed August 19, 2015); Judit Polgár, "Biography," Judit Polgár website, 2015, http://www.juditpolgar.com/en/biography (accessed August 19, 2015).

182 *games of that tournament—of 2735:* "Chessmetrics player profile: Sofia Polgar," at Chessmetrics, http://chessmetrics.com/cm/CM2/PlayerProfile.asp?Params=199510SSSSS1S102714000000111022676000246101oo (accessed August 20, 2015). See also "Zsofia Polgar," at Chessgames.com, http://www.chessgames.com/player/zsofia-polgar (accessed August 20, 2015).

She became a grandmaster: Myers, "Trained to be a genius."

184 *In a magazine interview:* Nancy Ruhling, "Putting a chess piece in the hand

of every child in America," *Lifestyles* (2006), reprinted in *Chess Daily News,* https://chessdailynews.com/putting-a-chess-piece-in-the-hand-of-every-child-in-america-2/ (accessed August 20, 2015).

a project at the University of Chicago: Benjamin S. Bloom, ed., *Developing Talent in Young People* (New York: Ballantine Books, 1985), 3–18.

185 *In the first stage:* Benjamin S. Bloom, "Generalizations about talent development," in ibid., 507–549.

186 *the three of them would regularly go down:* Matt Christopher and Glenn Stout, *On the Ice with . . . Mario Lemieux* (New York: Little, Brown, 2002).

turned many of his childhood activities into competitions: David Hemery, *Another Hurdle* (London: Heinemann, 1976), 9.

188 *the typical next step:* Bloom, "Generalizations about talent development," 512–518.

191 *"self-fueling, self-motivating drive for tremendous work":* David Pariser, "Conceptions of children's artistic giftedness from modern and postmodern perspectives," *Journal of Aesthetic Education* 31, no. 4 (1997): 35–47.

193 *how much it cost a family:* Kara Brandeisky, "What it costs to raise a Wimbledon champion," *Money,* July 4, 2014, http://time.com/money/2951543/cost-to-raise-tennis-champion-wimbledon/ (accessed August 23, 2015).

195 *people can train effectively well into their eighties:* K. Anders Ericsson, "How experts attain and maintain superior performance: Implications for the enhancement of skilled performance in older individuals," *Journal of Aging and Physical Activity* 8 (2000): 366–372.

the performance of master athletes has improved: Amanda Akkari, Daniel Machin, and Hirofumi Tanaka, "Greater progression of athletic performance in older Masters athletes," *Age and Ageing* 44, no. 4 (2015): 683–686.

a quarter of marathon runners in their sixties: Dieter Leyk, Thomas Rüther, Max Wunderlich, Alexander Sievert, Dieter Eßfeld, Alexander Witzki, Oliver Erley, Gerd Küchmeister, Claus Piekarski, and Herbert Löllgen, "Physical performance in middle age and old age: Good news for our sedentary and aging society," *Deutsches Aerzteblatt International* 107 (2010): 809–816.

the first person 100 years old or older: Karen Crouse, "100 years old. 5 world records," *New York Times,* September 21, 2015, http://www.nytimes.com/2015/09/22/sports/a-bolt-from-the-past-don-pellmann-at-100-is-still-breaking-records.html?module=CloseSlideshow®ion=SlideShowTopBar&version=SlideCard-10&action=click&contentCollection=Sports&pgtype=imageslideshow (accessed October 1, 2015).

196 *if ballet dancers are to develop the classic turnout:* Edward H. Miller, John N. Callander, S. Michael Lawhon, and G. James Sammarco, "Orthopedics and the classical ballet dancer," *Contemporary Orthopedics* 8 (1984): 72–97.

same sort of thing is true for the shoulders: John M. Tokish, "Acquired and adaptive changes in the throwing athlete: Implications on the disabled throwing shoulder," *Sports Medicine and Arthroscopy Review* 22, no. 2 (2014): 88–93.

The bones in a tennis player's dominant arm: Heidi Haapasalo, Saija Kontulainen, Hau Sievänen, Pekka Kannus, Markku Järvinen, and Ilkka Vuori, "Exercise-induced bone gain is due to enlargement in bone size without a change in volumetric bone density: A peripheral quantitative computed tomography study of the upper arms of male tennis players," *Bone* 27, no. 3 (2000): 351–357.

even tennis players who start later in life: Saija Kontulainen, Harri Sievänen, Pekka Kannus, Matti Pasanen, and Ilkka Vuori, "Effect of long-term impact-loading on mass, size, and estimated strength of humerus and radius of female racquet-sports players: A peripheral quantitative computed tomography study between young and old starters and controls," *Journal of Bone and Mineral Research* 17, no. 12 (2002): 2281–2289.

197 *Researchers have found proof of this, for example:* Gottfried Schlaug, Lutz Jäncke, Yanxiong Huang, Jochen F. Staiger, and Helmuth Steinmetz, "Increased corpus-callosum size in musicians," *Neuropsychologia* 33 (1995): 1047–1055.

a number of other regions of the brain: Dawn L. Merrett, Isabelle Peretz, and Sarah J. Wilson, "Moderating variables of music training—induced neuroplasticity: A review and discussion," *Frontiers in Psychology* 4 (2013): 606.

who started music training later and those who started earlier: Siobhan Hutchinson, Leslie Hui-Lin Lee, Nadine Gaab, and Gottfried Schlaug, "Cerebellar volume of musicians," *Cerebral Cortex* 13 (2003): 943–949.

198 *people who speak two or more languages:* Andrea Mechelli, Jenny T. Crinion, Uta Noppeney, John O'Doherty, John Ashburner, Richard S. Frackowiak, and Cathy J. Price, "Structural plasticity in the bilingual brain: Proficiency in a second language and age at acquisition affect grey-matter density," *Nature* 431 (2004): 757.

a study of multilingual people: Stefan Elmer, Jürgen Hänggi, and Lutz Jäncke, "Processing demands upon cognitive, linguistic, and articulatory functions promote grey matter plasticity in the adult multilingual brain: Insights from simultaneous interpreters," *Cortex* 54 (2014): 179–189.

199 *a quixotic undertaking:* Paul T. Brady, "Fixed-scale mechanism of perfect pitch," *Journal of the Acoustical Society of America* 48, no. 4, pt. 2 (1970): 883–887.

200 *a paper that described a training technique:* Lola L. Cuddy, "Practice effects in the absolute judgment of pitch," *Journal of the Acoustical Society of America* 43 (1968): 1069–1076.

201 *Mark Alan Rush, set out to test empirically:* Mark Alan Rush, "An experimental investigation of the effectiveness of training on absolute pitch in adult musicians" (Ph.D. diss., Ohio State University, 1989).

202 *a New Zealander named Nigel Richards:* The details on Nigel Richards come from several places. One good source is Stefan Fatsis, *Word Freak: Heartbreak, Triumph, Genius, on Obsession in the World of Competitive Scrabble* (New York: Houghton Mifflin Harcourt, 2001). See also Stefan Fatsis, "An outtake from *Word Freak:* The enigmatic Nigel Richards," *The Last Word* 21 (September 2011): 35–37, http://www.thelastwordnewsletter.com/Last_Word/Archives_files/TLW%20September%202011.pdf (accessed August 21, 2015); Oliver Roeder, "What makes Nigel Richards the best Scrabble player on earth," FiveThirtyEight, August 8, 2014, http://fivethirtyeight.com/features/what-makes-nigel-richards-the-best-scrabble-player-on-earth/ (accessed August 21, 2015).

203 *he won the 2015 French Scrabble championship:* Kim Willsher, "The French Scrabble champion who doesn't speak French," *The Guardian,* July 21, 2015, www.theguardian.com/lifeandstyle/2015/jul/21/new-french-scrabble-champion-nigel-richards-doesnt-speak-french (accessed August 21, 2015).

204 *Having studied many examples of creative genius:* Most of the thoughts on creative genius here can be found in K. Anders Ericsson, "Creative genius: A view from the expert performance approach," in *The Wiley Handbook of Genius,* ed. Dean Keith Simonton (New York: John Wiley, 2014), 321–349.

205 *a study of Nobel Prize winners:* Harriett Zuckerman, *Scientific Elite: Nobel Laureates in the United States* (New York: Free Press, 1977).

Chapter 8: But What About Natural Talent?

208 *the story that got told:* A quick Internet search will uncover a number of versions of the story. E.g., David Nelson, "Paganini: How the great violinist was helped by a rare medical condition," *News and Record* (Greensboro, NC), January 9, 2011, http://inmozartsfootsteps.com/1032/paganini-violinist-helped-by-marfan-syndrome/ (accessed August 21, 2015); "Nicolo Paganini," Paganini on the Web, http://www.paganini.com/nicolo/nicindex.htm (accessed August 21, 2015); "One string . . . and Paganini," Dr. S. Jayabarathi's Visvacomplex website, http://www.visvacomplex.com/One_String_and_Paganini.html (accessed August 21, 2015).

209 *he was truly a groundbreaking violinist:* See, for example, Maiko Kawabata, "Virtuosity, the violin, and the devil . . . What *really* made Paganini 'demonic'?" *Current Musicology* 83 (2007): 7–30.

an old scientific report: Edgar Istel and Theodore Baker, "The secret of Paganini's technique," *Musical Quarterly* 16, no. 1 (1930): 101–116.

210 *"Now the strings had to chide":* Ibid., 103.

212 *his achievements seem much less wondrous:* Andreas C. Lehmann and K. Anders Ericsson, "The historical development of domains of expertise:

Performance standards and innovations in music," in *Genius and the Mind: Studies of Creativity and Temperament in the Historical Record,* ed. Andrew Steptoe (Oxford: Oxford University Press, 1998), 64–97.

213 *According to many biographies:* There are many Mozart biographies. One particularly useful one, because it consists of accounts written during his lifetime, is Otto Erich Deutsch, *Mozart: A Documentary Biography,* 3rd ed. (London: Simon & Schuster, 1990). See also Edward Holmes, *The Life of Mozart* (New York: Cosimo Classics), 2005.

the piano concertos that Wolfgang "composed" at eleven: Jin Young Park, "A reinvestigation of early Mozart: The three keyboard concertos, K. 107" (Ph.D. diss., University of Oklahoma, 2002). See also Arthur Hutchings, *A Companion to Mozart's Piano Concertos* (Oxford, UK: Clarendon Press, 1999) and Wolfgang Plath, "Beitrâge zur Mozart-Autographie 1: Die Handschrift Leopold Mozarts" [The handwriting of Leopold Mozart], in *Mozart-Jahrbuch 1960/1961* (Salzburg: Internationalen Stiftung Mozarteum, 1961), 82–117.

214 *with every child prodigy I have looked into:* See more details on the Mario Lemieux story in K. Anders Ericsson, "My exploration for Gagné's 'evidence' for innate talent: It is Gagné who is omitting troublesome information so as to present more convincing accusations," in *The Complexity of Greatness: Beyond Talent or Practice,* ed. Scott Barry Kaufmann (New York: Oxford University Press, 2012), 223–256.

how the young Mario took to the ice: M. Brender, "The roots of Route 66," *Hockey News* (May 16 supplement: "Mario Lemieux's journey to greatness") 50, no. 35 (1997): 14.

some to claim that Lemieux is an example: François Gagné, "Yes, giftedness (aka 'innate' talent) does exist!" in Kaufmann, *Complexity of Greatness,* 191–222.

a little digging into Lemieux's childhood: Matt Christopher and Glenn Stout, *On the Ice with . . . Mario Lemieux* (New York: Little, Brown, 2002).

215 *the high jumper Donald Thomas:* David Epstein, *The Sports Gene: Inside the Science of Extraordinary Athletic Performance* (New York: Current, 2013). One example, among many, of where Epstein's story of Donald Thomas was featured is Tony Manfred, "This anecdote about high jumpers will destroy your faith in Malcolm Gladwell's 10,000-hours rule," *Business Insider,* August 15, 2013, http://www.businessinsider.com/high-jumpers-anecdote-questions-gladwells-10000-hours-rule-2013-8 (accessed August 21, 2015).

Here are the basics: USTFCCCA (U.S. Track & Field and Cross Country Coaches Association), "USTFCCCA profile of Donald Thomas: An improbable leap into the limelight," *Track and Field News,* http://trackandfieldnews.com/index.php/display-article?arId=15342 (accessed August 21, 2015).

217 *"something like 6-2, 6-4, nothing memorable":* Ibid.

218 *the ability to jump off of one leg:* Guillaume Laffaye, "Fosbury Flop:

Predicting performance with a three-variable model," *Journal of Strength & Conditioning Research* 25, no. 8 (2011): 2143–2150.

220 *people with savant syndrome:* A special issue of the *Philosophical Transactions of the Royal Society B* is entirely devoted to the savant syndrome and, in particular, its relationship with autism, and it is a good source for current thinking on the savant syndrome. See, in particular, the overview article, Darold A. Treffert, "The savant syndrome: An extraordinary condition. A synopsis: Past, present, and future," *Philosophical Transactions of the Royal Society B* 364, no. 1522 (2009): 1351–1357.

they have worked for it, just like anyone else: A good general-audience review of new thinking on the savant syndrome is Celeste Biever, "The makings of a savant," *New Scientist* 202, no. 2711 (June 6, 2009): 30.

autistic savants are much more likely than the nonsavants: Francesca Happé and Pedro Vital, "What aspects of autism predispose to talent?" *Philosophical Transactions of the Royal Society B* 364, no. 1522 (2009): 1369–1375.

221 *Donny is addicted to dates:* Jennifer Vegas, "Autistic savant 'addicted' to dates," *ABC Science,* January 31, 2007, http://www.abc.net.au/science/articles/2007/01/31/1837037.htm (accessed June 26, 2015).

He has memorized all fourteen possible yearly calendars: Marc Thioux, David E. Stark, Cheryl Klaiman, and Robert T. Schultz, "The day of the week when you were born in 700 ms: Calendar computation in an autistic savant," *Journal of Experimental Psychology: Human Perception and Performance* 32, no. 5 (2006): 1155–1168.

a psychologist named Barnett Addis: Barnett Addis, "Resistance to parsimony: The evolution of a system for explaining the calendar-calculating abilities for idiot savant twins" (paper presented at the meeting of the Southwestern Psychological Association, New Orleans, April 1968). For more details on the twins, see O. A. Parsons, "July 19, 132,470 is a Saturday: Idiot savant calendar-calculating twins" (paper presented at the meeting of the Southwestern Psychological Association, New Orleans, April 1968).

222 *noted in a 1988 review:* K. Anders Ericsson and Irene Faivre, "What's exceptional about exceptional abilities?" In *The Exceptional Brain: Neuropsychology of Talent and Special Abilities,* ed. Loraine K. Obler and Deborah Fein (New York: Guilford, 1988), 436–473.

More recent case studies: See, for example, G. L. Wallace, F. Happé, and J. N. Giedd, "A case study of a multiply talented savant with an autism spectrum disorder: Neuropsychological functioning and brain morphometry," *Philosophical Transactions of the Royal Society of London Series B, Biological Sciences* 364 (2009): 1425–1432; and Richard Cowan and Chris Frith, "Do calendrical savants use calculation to answer date questions? A functional magnetic resonance imaging study," *Philosophical Transactions of the Royal Society of London Series B, Biological Sciences* 364 (2009): 1417–1424.

223 *adults believe they can't sing:* Lola L. Cuddy, Laura-Lee Balkwill, Isabelle

Peretz, and Ronald R. Holden, "Musical difficulties are rare: A study of 'tone deafness' among university students," *Annals of the New York Academy of Sciences* 1060 (2005): 311–324.

these people aren't very happy about it: Susan Knight, "Exploring a cultural myth: What adult non-singers may reveal about the nature of singing," *Phenomenon of Singing* 2 (2013): 144–154.

someone convinced them: Ibid.

224 *a woman with this condition:* Isabelle Peretz, Julie Ayotte, Robert J. Zatorre, Jacques Mehler, Pierre Ahad, Virginia B. Penhune, and Benoît Jutras, "Congenital amusia: A disorder of fine-grained pitch discrimination," *Neuron* 33 (2002): 185–191.

no evidence that large numbers of people: Magdalena Berkowska and Simona Dalla Bella, "Acquired and congenital disorders of sung performance: A review," *Advances in Cognitive Psychology* 5 (2009): 69–83; Karen J. Wise and John A. Sloboda, "Establishing an empirical profile of self-defined 'tone deafness': Perception, singing performance and self-assessment," *Musicae Scientiae* 12, no. 1 (2008): 3–26. See also Knight, "Exploring a cultural myth."

there are some cultures: Knight, "Exploring a cultural myth."

a curriculum called Jump Math: David Bornstein, "A better way to teach math," *New York Times,* April 11, 2011, http://opinionator.blogs.nytimes.com/2011/04/18/a-better-way-to-teach-math/?_r=0 (accessed August 21, 2015).

227 *Some of the earliest work was done in the 1890s:* Alfred Binet, *Psychologie des grands calculateurs et joueurs d'echecs* [The psychology of great calculators and chess players] (Paris: Libraire Hachette, 1894).

One of the most enlightening of these studies: Merim Bilalić, Peter McLeod, and Fernand Gobet, "Does chess need intelligence? A study with young chess players," *Intelligence* 35 (2007): 457–470.

228 *a relationship between IQ and chess-playing ability:* Dianne D. Horgan and David Morgan, "Chess expertise in children," *Applied Cognitive Psychology* 4 (1990): 109–128; Marcel Frydman and Richard Lynn, "The general intelligence and spatial abilities of gifted young Belgian chess players," *British Journal of Psychology* 83 (1992): 233–235.

no better visuospatial abilities: See, for instance, Andrew J. Waters, Fernand Gobet, and Gerv Leyden, "Visuo-spatial abilities in chess players," *British Journal of Psychology* 93 (2002): 557–565; Josef M. Unterrainer, Christoph P. Kaller, Ulrike Halsband, and B. Rahm, "Planning abilities and chess: A comparison of chess and non-chess players on the Tower of London," *British Journal of Psychology* 97 (2006): 299–311; Roland H. Grabner, Aljoscha C. Neubauer, and Elbeth Stern, "Superior performance and neural efficiency: The impact of intelligence and expertise," *Brain Research Bulletin* 69 (2006): 422–439; and Jörg Doll and Ulrich Mayr, "Intelligenz und Schachleistung—eine Untersuchung an Schachexperten" [Intelligence and chess performance — A study of chess experts], *Psychologische Beiträge* 29 (1987): 270–289. An early study of grandmasters can be found in I. N. Djakow, N. W. Petrowski, and P. A. Rudik, *Psychologie des Schachspiels* [Psychology of chess playing] (Berlin: de Gruyter, 1927).

do not have systematically higher IQs: Josef M. Unterrainer, Christoph P. Kaller, Ulrike Halsband, and B. Rahm, "Planning abilities and chess: A comparison of chess and non-chess players on the Tower of London," *British Journal of Psychology* 97 (2006): 299–311; Roland H. Grabner, Aljoscha C. Neubauer, and Elbeth Stern, "Superior performance and neural efficiency: The impact of intelligence and expertise," *Brain Research Bulletin* 69 (2006): 422–439.

no correlation between the IQs: Jörg Doll and Ulrich Mayr, "Intelligenz und Schachleistung — eine Untersuchung an Schachexperten" [Intelligence and chess performance — A study of chess experts], *Psychologische Beiträge* 29 (1987): 270–289.

229 *Recent studies of Go masters:* Boreom Lee, Ji-Young Park, Wi Hoon Jung, Hee Sun Kim, Jungsu S. Oh, Chi-Hoon Choi, Joon Hwan Jang, Do-Hyung Kang, and Jun Soo Kwon, "White matter neuroplastic changes in long-term trained players of the game of 'Baduk' (GO): A voxel-based diffusion-tensor imaging study," *NeuroImage* 52 (2010): 9–19; Wi Hoon Jung, Sung Nyun Kim, Tae Young Lee, Joon Hwan Jang, Chi-Hoon Choi, Do-Hyung Kang, and Jun Soo Kwon, "Exploring the brains of *Baduk* (Go) experts: Gray matter morphometry, resting-state functional connectivity, and graph theoretical analysis," *Frontiers in Human Neuroscience* 7, no. 633 (2013): 1–16.

score no higher on IQ tests: Because people who score higher on IQ tests are more likely to do well in school and also to stay in school — a phenomenon that has been repeatedly observed — it is possible that some young Go players with lower IQs quit school earlier than their peers in order to fully focus on studying Go. This could explain why professional Go players have IQs that were lower than average.

233 *evidence of this pattern in many different fields:* For a review with a long list of references to various studies, see K. Anders Ericsson, "Why expert performance is special and cannot be extrapolated from studies of performance in the general population: A response to criticisms," *Intelligence* 45 (2014): 81–103.

a study of ninety-one fifth-grade students: William T. Young, "The role of musical aptitude, intelligence, and academic achievement in predicting the musical attainment of elementary instrumental music students," *Journal of Research in Music Education* 19 (1971): 385–398.

234 *no relationship between IQ and music performance:* Joanne Ruthsatz, Douglas Detterman, William S. Griscom, and Britney A. Cirullo, "Becoming an expert in the musical domain: It takes more than just practice," *Intelligence* 36 (2008): 330–338.

a study on expertise in oral surgery: Kyle R. Wanzel, Stanley J. Hamstra, Marco F. Caminiti, Dimitri J. Anastakis, Ethan D. Grober, and Richard K. Reznick, "Visual-spatial ability correlates with efficiency of hand motion and successful surgical performance," *Surgery* 134 (2003): 750–757.

people studying to be London taxi drivers: Katherine Woollett and Eleanor A. Maguire, "Acquiring 'the knowledge' of London's layout drives structural brain changes," *Current Biology* 21 (2011): 2109–2114.

no correlation between IQ and scientific productivity: Robert S. Root-Bernstein, Maurine Bernstein, and Helen Garnier, "Identification of scientists making long-term, high impact contributions, with notes on their methods of working," *Creativity Research Journal* 6 (1993): 329–343; Kenneth S. Law, Chi-Sum Wong, Guo-Hua Huang, and Xiaoxuan Li, "The effects of emotional intelligence on job performance and life satisfaction for the research and development scientists in China," *Asia Pacific Journal of Management* 25 (2008): 51–69.

had an IQ of 125: For information on Feynman, Watson, and Shockley, see Robert Root-Bernstein, Lindsay Allen, Leighanna Beach, Ragini Bhadula, Justin Fast, Chelsea Hosey, Benjamin Kremkow, Jacqueline Lapp, Kaitlin Lonc, Kendell Pawelec, Abigail Podufaly, Caitlin Russ, Laurie Tennant, Eric Vrtis, and Stacey Weinlander, "Arts foster scientific success: Avocations of Nobel, National Academy, Royal Society, and Sigma Xi members," *Journal of the Psychology of Science and Technology* 1, no. 2 (2008): 51–63.

235 *some fields need an IQ score of around 120:* Donald W. MacKinnon, "The nature and nurture of creative talent," *American Psychologist* 17, no. 7 (1962): 484–495.

236 *A 2012 study of tennis players:* Jessie Brouwers, Veerle de Bosscher, and Popi Sotiriadou, "An examination of the importance of performances in youth and junior competition as an indicator of later success in tennis," *Sport Management Review* 15 (2012): 461–475.

237 *children with a temperament that encourages social interaction:* Melanie Noel, Carole Peterson, and Beulah Jesso, "The relationship of parenting stress and child temperament to language development among economically disadvantages preschoolers," *Journal of Child Language* 35, no. 4 (2008): 823–843.

infants who paid more attention to a parent: Brad M. Farrant and Stephen R. Zubrick, "Parent-child book reading across early childhood and child vocabulary in the early school years: Findings from the Longitudinal Study of Australian Children," *First Language* 33 (2013): 280–293.

239 *a story in his book* Outliers*:* Malcolm Gladwell, *Outliers: The Story of Success* (New York: Little, Brown, 2008).

240 *advantage among hockey players does seem to taper off:* See, for example, Benjamin G. Gibbs, Mikaela Dufur, Shawn Meiners, and David Jeter, "Gladwell's big kid bias?" *Contexts* 9, no. 4 (2010): 61–62.

241 *experience playing linear board games:* Robert S. Siegler and Geetha B. Ramani, "Playing board games promotes low-income children's numerical development," *Developmental Science* 11 (2008): 655–661.

Chapter 9: Where Do We Go from Here?

243 *This glimpse came courtesy of three researchers:* Louis Deslauriers, Ellen Schelew, and Carl Wieman, "Improved learning in a large-enrollment physics class," *Science* 332 (2011): 862–864.

245 *get them to practice thinking like physicists:* Ibid. Also see Jeffrey Mervis, "Transformation is possible if a university really cares," *Science* 340, no. 6130 (2013): 292–296.

247 *For the sake of comparison:* Deslauriers, Schelew, and Wieman, "Improved learning."

249 *I have also been working with Rod Havriluk:* See the website of Havriluk's company, Swimming Technology Research: https://swimmingtechnology.com/.

250 *The first thing that Wieman and his colleagues did:* Deslauriers, Schelew, and Wieman, "Improved learning."

251 *the crucial ingredient:* David Bornstein, "A better way to teach math," *New York Times,* April 11, 2011, http://opinionator.blogs.nytimes.com/2011/04/18/a-better-way-to-teach-math/?_r=0 (accessed August 21, 2015).

252 *Research comparing physics experts with physics students:* R. R. Hake, "Interactive-engagement vs. traditional methods: A six-thousand student survey of mechanics test data for introductory physics students," *American Journal of Physics* 66, no. 4 (1998): 64–74; David Hestenes, Malcolm Wells, and Gregg Swackhamer, "Force concept inventory," *Physics Teacher* 30 (1992): 141–158.

cannot correctly explain what causes the changing seasons: Eve Kikas, "Teachers' conceptions and misconceptions concerning three natural phenomena," *Journal of Research in Science Teaching* 41 (2004): 432–448; Yaël Nazé and Sebastien Fontaine, "An astronomical survey conducted in Belgium," *Physics Education* 49 (2014): 151–163.

An amusing video taken at a Harvard University commencement: "Harvard graduates explain seasons," YouTube, https://www.youtube.com/watch?v=powk4qG2mIg (accessed October 4, 2015).

253 *Wieman and his colleagues pretested the clicker questions:* Deslauriers, Schelew, and Wieman, "Improved learning."

254 *deliberate-practice methods were adopted:* Jeffrey Mervis, "Transformation is possible if a university really cares," *Science* 340, no. 6130 (2013): 292–296.

257 *the psychological state of "flow":* Mihaly Csikszentmihalyi, *Flow: The Psychology of Optimal Experience* (New York: Harper & Row, 1990).

Index

penguin.co.uk/vintage